For further volumes:
http://www.springer.com/series/5584

Eleftherios Mylonakis · Frederick M. Ausubel
Michael Gilmore · Arturo Casadevall
Editors

Recent Advances on Model Hosts

Springer

Editors
Eleftherios Mylonakis
Harvard Medical School
Massachusetts General Hospital
Boston, MA, USA
emylonakis@partners.org

Frederick M. Ausubel
Harvard Medical School
Massachusetts General Hospital
Boston, MA, USA
ausubel@molbio.mgh.harvard.edu

Michael Gilmore
Harvard Medical School
Schepens Eye Research Institute
Boston, MA, USA
mgilmore@vision.eri.harvard.edu

Arturo Casadevall
Department of Microbiology
and Immunology
Albert Einstein College of Medicine
Yeshiva University of Medicine
Bronx, NY, USA
arturo.casadevall@einstein.yu.edu

ISSN 0065-2598
ISBN 978-1-4419-5637-8 e-ISBN 978-1-4419-5638-5
DOI 10.1007/978-1-4419-5638-5
Springer New York Dordrecht Heidelberg London

Library of Congress Control Number: 2011939199

Printed on acid-free paper

Springer is part of Springer Science+Business Media (www.springer.com)

Contents

Chapter 1
Amoeba Provide Insight into the Origin of Virulence in Pathogenic Fungi

Arturo Casadevall

Abstract Why are some fungi pathogenic while the majority poses no threat to humans or other hosts? Of the more than 1.5 million fungal species only about 150–300 are pathogenic for humans, and of these, only 10–15 are relatively common pathogens. In contrast, fungi are major pathogens for plants and insects. These facts pose several fundamental questions including the mechanisms responsible for the origin of virulence among the few pathogenic species and the high resistance of mammals to fungal diseases. This essay explores the origin of virulences among environmental fungi with no obvious requirement for animal association and proposes that selection pressures by amoeboid predators led to the emergence of traits that can also promote survival in mammalian hosts. In this regard, analysis of the interactions between the human pathogenic funges *Cryptococcus neoformans* and amoeba have shown a remarkable similarity with the interaction of this fungus with macrophages. Hence the virulence of environmental pathogenic fungi is proposed to originate from a combination of selection by amoeboid predators and perhaps other soil organism with thermal tolerance sufficient to allow survival in mammalian hosts.

The Pathogenic Fungi

The human pathogenic fungi comprise a highly diverse group of organism that can be broadly classified into two broad groups: dermatophytes and systemic mycoses. The dermatophytes are relatively common pathogens and include such agents as Tinea pedis, a cause of athlete's feet. Dermatophytes cause troublesome conditions but are rarely life-threatening. In contrast, systemic mycoses are rare in immunologically intact human populations. In comparison to bacterial and viral diseases that are known since antiquity, systemic appear to be relative latecomers to the parade of human pathogens. Diseases such as cryptococcosis, blastomycosis, histoplasmosis and coccidiomycosis were only described in the late nineteenth or early twentieth centuries. In fact, most pathogenic fungal species were identified in the twentieth century and the overwhelming majority is exceedingly rare, thus related to case reports. Although sporadic systemic fungal diseases almost certainly

A. Casadevall (✉)
Departments of Microbiology and Immunology and Medicine (Division of Infectious Diseases), Albert Einstein College of Medicine, 1300 Morris Park Ave., Bronx, NY 10461, USA
e-mail: arturo.casadevall@einstein.yu.edu

E. Mylonakis et al. (eds.), *Recent Advances on Model Hosts*,
Advances in Experimental Medicine and Biology 710, DOI 10.1007/978-1-4419-5638-5_1,

must have affected certain individuals in human populations since the emergence of humanity, these were not recognized as such, probably because of the relatively paucity of cases and the fact that their non-contagiousness meant the absence of epidemics with recognizable common symptoms that would have led to their identification as cause of disease.

Endothermy and Susceptibility to Fungal Diseases

The paucity of fungal diseases in mammals relative to other classes of pathogenic microbes is striking given that fungi are major pathogens for plants, insects and amphibians. In fact, fungi are the major pathogens of plants and a chytrid fungus has caused the extinction of numerous amphibian species in recent years. Given the susceptibility of amphibians the relative resistance of mammals to fungi cannot be attributed to be a general characteristic of vertebrates. When considering the mammals relative to other vertebrates and animals they have several striking characteristics such as being endothermic, homeothermic and possession an immune system capable of both innate and adaptive immunity. The same characteristics also apply to birds, which are also relative resistant to fungal infections except for thermophilic fungal genera such as *Aspergillus* spp.

Evidence for the synergistic contribution of temperature and sophisticated immunity to host resistance against fungi is provided by the work of Dr. John Perfect using rabbits as a model for cryptococcosis (Perfect et al. 1980). Rabbits have the relatively high body temperature for mammals with core temperatures of 41°C, a temperature that is near the temperature tolerance limit for many cryptococcal strains. Hence, it is very difficult to induce systemic cryptococcosis in rabbits and this presumably is a consequence of the fact that fungi near their thermal limit of viability are in such stressed conditions that they are highly vulnerable to host defense. However, localized infection can be caused in rabbits by injecting yeast cells into testis, an organ that is several degrees cooler than the core, implying that the immune system alone cannot clear the infection when the fungus is in a less thermally stressed condition. However, when rabbits are immunosuppressed with corticosteroids it is possible to induce meningoencephalitis by direct injection of cryptococci into the cerebrospinal fluid, such that even thermally stressed yeast cells can persist in tissue in the setting of impaired immunity.

Additional evidence for the importance of the contribution of endothermy to mammalian host defense against fungi comes from the ongoing decline of bats in North America as a consequence of a new fungal disease caused by Geomyces destructans (Blehert et al. 2009). Beginning in 2006 bat die offs were reported in the Northeast of the United States accompanied by fungal growth in the bat nostrils in a disease named 'white nose syndrome'. *Geomyces destructans* is a cold loving fungus that grows optimally at 12°C. During the summers bats are awake and maintain mammalian range temperatures but they hibernate during winters and their core temperature drops to the 6–11°C range, a permissive temperature for *Geomyces destructants* (Chaturvedi et al. 2010). Core temperatures have been shown to be important in both susceptibility and resistance to fungal diseases even among ectothermic vertebrates such as reptiles and amphibians. Lizards are unable to survive experimental fungal infection unless they can induce behavioral fevers. Similarly, frogs can be cured of chytrid infection when placed at 37°C.

Analysis of the thermal tolerance of several thousand fungal strains in a culture collection revealed that most thrived at ambient temperatures while undergoing a rapid decline in their growth viability at temperatures above 30°C (Robert and Casadevall 2009). A plot of the proportion of viable fungal strains versus temperature revealed a linear relationship such that for each degree in temperature increase between 30°C and 42°C there was a corresponding decline of 6% in terms strain viability (Robert and Casadevall 2009). Consequently, mammalian temperatures were sufficient to inhibit the majority of fungal strains. However, the mammalian lifestyle is characterized by a high basal metabolic rate that demands a high food energy intake. There is no good explanation for the high temperatures associated with mammalian endothermy and homeothermy. However, solving for the temperature that optimally restricts fungal growth and metabolic costs yielded a temperature of 37°C, providing the tantalizing suggestion that the costly mammalian lifestyle may have been selected for the protection it provides against fungal diseases (Bergman and Casadevall 2010). If the temperature gradient between mammalian temperatures and the environment provides protection against environmental pathogens there is concern that global warming will reduce it and bring forth new fungal diseases (Garcia-Solache and Casadevall 2010).

Origin of Fungal Infection

When considering the virulence of pathogenic fungi it is instructive to categorize them by the place from which fungal infection is contracted. There are two major sources of fungal infection: other hosts and the environment. Humans are born sterile but are then rapidly infected with a large number of microbes that eventually becomes the commensal or host-associated microbiota. In this regard, Candida albicans is often acquired from the maternal flora and dermatophytes are probably acquired from other infected hosts. In contrast, most organisms responsible for systemic mycoses are probably acquired directly from environmental reservoirs. For example, a study of the age of seroconversion for cryptococcal infection of children from New York City revealed that the majority acquired serological evidence of infection in early life (Goldman et al. 2001), presumably by infection from urban sources of C. *neoformans* (Currie et al. 1994). Similarly, such organisms as *Blastomyces dermatitidis*, *Cocciodiodes immitis*, and *Histoplasma capsulatum* are likely to be acquired directly from environmental sources.

In general, human disease from fungi acquired from other hosts usually results from an alteration of the host–microbe relationship such that tissue damage affects homeostasis. For example, mucosal and/or systemic candidiasis is often a consequence of immune suppression, alterations in host bacterial flora and comprised integument. In recent years evidence has accumulated suggesting that some cases of chronic candidiasis are associated with genetic traits, such as mutations in the dectin signaling pathways. Such mutations are believed to impair host responses and thus alter the normal balance tipping the outcome towards disease. Similarly, dermatophyte-related disease such as athlete's foot occurs when the pedal environment is disturbed by shoe usage thus creating conditions for fungal proliferation and skin damage. In contrast, human disease from fungi acquired from the environment is usually self limited with severe cases reflecting infection in hosts with impaired immunity or large innocula as are sometimes encountered in special situations such as histoplasmosis following cave explorations and coccidiomycosis following paleontological excavations.

Fungi acquired from other hosts such as *Candida albicans* are already host adapted and they do not cause disease in the majority of infected hosts. In contrast, when fungi acquired from the environment infect a mammalian host they must adapt and survive in an environment that is very different than their natural habitat. Hence, the origin of virulence for fungi acquired from other hosts or directly from the environment must have fundamental differences with regards to fungal adaptation to the host and microbial traits necessary for survival in the host. In the case of *C. albicans* there is conclusive evidence that the same strains that exist in a commensal state in one individual can be associated with disease in the same individual with there is an alteration of the host–microbe relationship. Consequently, the same organism is pathogenic or non-pathogenic depending on the host state. However, for those organisms acquired directly from the environment inhabiting a mammal is not part of their life cycle and carries considerable danger of lethality from elevated temperatures and powerful immune defenses. Hence, the few environmental fungi capable of causing human disease must be endowed with traits that allow survival in mammalian hosts, and these traits must be independent of their ability to survive in such hosts given than neither animal virulence or carriage can be considered essential to their survival in the soil.

The Soil Environment

The human pathogenic fungi acquired directly from the environment inhabit soils, but these soils can differ significantly. For example, *C. neoformans* and *H. capsulatum* are often found in soils contaminated with bird and bat guano, respectively, while *C. immitis* is found in desert soils of the North American southwest. Soils are remarkably complex ecologic niches occupied by a complex biota with which such fungi must interact. A fascinating aspect of these soil-dwelling fungi is that they can recovered from soils in a fully virulent form. For example, early techniques for the isolation of *C. neoformans* and *H. capsulatum* from soils included injecting a soil slurry into rodents, which cleared other soil microbes but developed infection with those specific organisms. Similarly, placement of rodents in soils known to be infected with *C. immitis* resulted in their acquiring coccidiodiomycosis. These observations are significant because they indicate that soil residence is associated with the maintenance of pathogenic potential for mammalian hosts. Consequently, there is a high likelihood that selection pressures in the environment are responsible for the emergence and maintenance of traits that confer upon some soil microbes the capacity for survival in animal hosts. In considering the types of interactions that could have selected for attributes of virulence we have focused on amoeboid predators for these inhabit soils and share certain characteristics with cells of the host innate immune system.

Amoeba and Fungi Coexist in Many Ecologic Sites

Amoeba and fungi are found throughout the terrestrial environment and their co-existence in many ecologic niches provides them with the opportunity to interact. Although amoebae are generally associated with watery and moist environments they have the capacity to survive in dry conditions as a result of encystment. Consequently, amoeba can be found even in arid soils (Robinson et al. 2002), and this is important because it implies that even desert-inhabiting species like *C. immitis* can come into contact with them. Some species of amoeba are thermotolerant such that they can inhabit hot springs (Sheehan et al. 2003). Amoeba are frequent on the surfaces and internal tissues of edible mushrooms (Napolitano 1982), attesting to a close connection between the kingdoms. Amoebae are also found in the human mouth and nasal mucosa (Rivera et al. 1984).

When considering fungal–amoeba interactions is it important to note that there is tremendous diversity among amoebae species and that the biological diversity for this group of organisms is poorly understood. Estimates of the number of amoebae species in the biosphere prior to the development of deep sequencing techniques ranged from 40,000 to 100,000, with the understanding that such numbers were likely to be underestimates (Couteaux and Darbyshire 1988). In general, amoeba feast on bacteria and their prevalence in an ecologic site is often directly proportional to the numbers of bacteria found (Ettinger et al. 2003). However, there some types of amoebae that specialize in eating fungi and these are known as mycophageous amoebae (Couteaux and Darbyshire 1988).

The majority of laboratory studies of fungal–amoeba interactions have focused on only a few species of amoebae such as *Acanthamoebae castellanii* and *polyphaga*. The extent to which these species provide information that is representative of the types of fungal–amoeba interactions found in the environment is not known. Of these two amoebae *A. castellani* provides the major system because it has been adapted to grow in an axenic medium, an adaptation that greatly simplifies experimental design by eliminating the need for feeder bacteria but also provides for a much more artificial type of situation. Hence, when considering results obtained in laboratory conditions it is important to remember that such systems are quite distant from natural systems and that there is little or no information about the generalizability of observations to other amoebae species.

When provided with a choice of food prey amoebae can be choosy and different amoebae have different culinary interests. For example when the soil amoebae *Hartmannella glebae* is offered a choice of menus it prefers gram positive bacteria over gram negative bacteria and ignores algae, yeasts and molds (Upadhyay 1968). On the other hand, *A. castellanii* was originally isolated discovered as a contaminant in cultures of a *Cryptococcus* sp., where it was probably feasting (Castellani 1931), a finding supported by subsequent investigators who established that it preyed on yeast (Nero et al. 1964). *A. castellani* recognizes fungal cells through a mannose-binding receptor and phagocytosis places ingested yeast in the lysosomal pathway, which produces nutrients and activates phosphoinositide metabolism (Allen and Dawidowicz 1990a, b. The presence of surface receptors presumably allows *A. castellani* to discriminate between yeast particles and latex beads, such that it preferentially ingests the former (Bowers and Olszewski 1983). There is indirect evidence that yeast cells can produce products that interfere with phagocytosis, possibly as a defense mechanism against ameboid predators. For example, co-incubation of *Entoamoeba histolytica* with *Saccharomyces boulardii* and erythrocytes reduces erythrocyte ingestion, suggesting that yeast cells produce soluble factors that reduce amoeba phagocytic capacity (Rigothier et al. 1994).

Amoebae as Model Hosts for the Study of Fungal Virulence

The fact that amoebae resemble host phagocytic cells such as macrophages, which are essential cells in containing fungal infection, together with knowledge that they interact with fungi in ecologic sites and in the laboratory suggested their potential usefulness for studying fungal virulence. Early studies on the interaction of amoebae with human pathogenic fungi were carried out in the 1950s when Castellani demonstrated that *A. castellani* could devour and kill a strain of *Cryptococcus neoformans* (Castellani 1955). Subsequently, Bulmer and colleagues followed up the interesting observation that attempts to culture *C. neoformans* from mouse feces were often complicated by contamination of recovery plates with amoeba (Neilson et al. 1978). They identified the contaminating amoeba as *A. polyphaga* and proceeded to demonstrate that this strain devoured yeast cryptococcal cells but avoided rare pseudohyphal forms and proposed that a morphological transition from yeast to pseudohyphal provided an 'escape hatch' for *C. neoformans* to survive amoeba predation (Neilson et al. 1978; Bunting et al. 1979). When pseudohyphal forms were tested for virulence in a mouse model of infection they were found to be significantly less virulent than yeast forms, establishing that amoeba predation could select for variants with altered virulence. Apart from establishing the precedent of altered virulence following predation, the relevance of pseudohyphal strains in infection is unclear since such forms are very rare with *C. neoformans* and do not seem to play a major role in pathogenesis. The observation that some amoeba preyed on *C. neoformans* led Bulmer and colleagues to propose that these protists were major biological control agents in the environment, a finding supported by the observation that cryptococcal numbers in soils were inversely proportional to the presence of amoeba (Neilson et al. 1978; Bunting et al. 1979; Ruiz et al. 1982).

After the studies of Bulmer work with amoeba ceased until the late 1990s when these hosts again became interesting objects for investigation as a result of progress in understanding *C. neoformans*–macrophage interactions. At that time it was becoming apparent, and increasingly accepted, that *C. neoformans* was a facultative intracellular pathogen in mammalian hosts (Feldmesser et al. 2000, 2001). Furthermore, it was clear that *C. neoformans* survived in intracellular spaces with a unique pathogenic strategy that included no interference with initial phagosome maturation followed by later phagosomal membrane leakiness and the release of a large number of polysaccharide-containing vesicles into the host cell cytoplasm. The recognition that *C. neoformans* was an intracellular pathogen with a sophisticated and unique strategy introduced a conundrum since it was not immediately clear why a soil organism with no requirement for mammalian hosts for survival of completion of its life cycle would possess such attributes. This observation combined with earlier findings that *C. neoformans* and amoeba interacted led to the hypothesis that such pathogenic strategies are the result of environmental selection for characteristics that accidentally led to traits that permitted survival in host phagocytic cells. The hypothesis was tested by investigating *C. neoformans* interactions with *A. castellani* and it was found that features of macrophage infection such as intracellular vesicular release were faithfully reproduced in amoebae (Steenbergen et al. 2001). Of note, *C. neoformans* attributes needed for mammalian virulence, such as the capsule, melanin synthesis and phospholipase were also found to be important for virulence in amoeba (Steenbergen et al. 2001). Those observations led to a formal proposal that virulence in *C. neoformans* and other environmental organisms with no requirements for animal hosts was a consequence of selection of traits that serendipitously also provided the means to survive animal immune defense mechanisms (Casadevall et al. 2003; Steenbergen and Casadevall 2003). In subsequent essays the proposal was expanded and modified to emphasize the point that such interactions may arise stochastically in the environment and result in the steady emergence of microbes with virulence potential for other hosts (Casadevall and Pirofski 2007; Casadevall 2007).

Subsequent studies have shown that not all virulence factors for mammals are necessary for virulence in amoebae. For example, the alpha mating factor locus is important has been shown to contribute to virulence in mice but no difference was observed in the interaction of congenic a and a strains in amobae (Nielsen et al. 2005). Similarly, the Ca^{2+}–ATPase (*ECA1*) type calcium pump of *C. neoformans* is involved in stress tolerance and is critical for thermolerance. *C. neoformans* strains deficient in ECA1 are avirulent for mice because of reduced thermotolerance but hypervirulent for A. castellani (Fan et al. 2007). The finding that some but not all virulence factors are required for animal and protist virulence can be expected from the enormous differences inherent in the comparison between these types of hosts. Certainly, amoeba have no immune systems and virulence factors that promote fitness in animal hosts by modulating immune function would not be expected to be relevant for amoeba pathogenesis. Similarly, fungal structures that affected amoeba receptors for yeast cell phagocytosis would not be relevant for mammalian virulence unless they had a similar impact on macrophages through the targeting of conserved pathways.

In an effort to broaden the scope of this concept our group also studied the interactions with amoeba and several other pathogenic fungi: *Candida albicans* (Steenbergen et al. 2001), *Blastomyces dermatitidis* (Steenbergen et al. 2004), *Sporothrix schenkii* (Steenbergen et al. 2004), and *H. capsulatum* (Steenbergen et al. 2004). *C. albicans* cells were killed by *A. castellani* while cells of the three dimorphic pathogenic fungi each survived the interaction with amoeba and killed their hosts. Cells of *S. schenkii* and *H. capsulatum* were readily ingested by amoeba but survived intracellularly while *B. dermatitidis* was seldom ingested. Of great interest was the serendipitous observation that co-incubation of amoebae and cells of the three dimorphic fungi led to an increase in hyphal forms even at 37°C (Steenbergen et al. 2004). This observation was important because the yeast-hyphal transition is associated with virulence in several dimorphic fungi and this transition occurs as the temperature is changed from ambient (25°C; hyphal form) to mammalian (37°C; yeast form). Given that the yeast morphology is the type most likely associated with virulence it was always assumed that this transition was somehow related to the capacity for mammalian virulence but the mechanisms for evolutionary emergence where unknown, since again, none of these fungi has a requirement for an animal host for survival. Hence, the finding that the hyphal form could be induced at the non-permissive temperature of 37°C by amoebae implied that this trait was also a consequence of environmental selection, possibly another form of the 'escape hatch' view first proposed by Bulmer when he noted that pseudohyphal forms of *C. neoformans* resisted predation (Neilson et al. 1978).

In contrast to *C. neoformans*, the interaction of *C. gattii* with *A. castellani* appears to be qualitatively different (Malliaris et al. 2004). Whereas, *C. neoformans* is readily ingested and appears to kill the amoeba as a consequence of intracellular pathogenesis, *C. gattii* is seldom found internalized and kills amoeba through an extracellular mechanism. The resistance of *C. gattii* to phagocytosis appeared related to differences in the capsular polysaccharide structure. Subsequent studies established that the capsule protected against ingestion by amoeba providing another correlate between the function of this structure in mammalian pathogenesis and interactions with amoebae (Chrisman et al. 2010; Zaragoza et al. 2008). Another important parallel between *C. neoformans* interactions with amoebae and macrophages was related to the phenomenon of exiting from phagocytic cells. *C. neoformans* has been shown to be capable of non-lytic exocytosis from murine and human macrophages in a remarkable phenomenon where the host cells survive the event (Ma et al. 2006; Alvarez and Casadevall 2006; Alvarez et al. 2009). This phenomenon involves a complex cellular choreography whereby phagosomes containing *C. neoformans* are expulsed from the macrophages, either before or after phagosome to phagosome fusion, by a mechanism that must involve phagosomal-cell membrane fusion in a manner that preserves the integrity of the host cells. Non-lytic exocytosis has now been described in amobae (Chrisman et al. 2010). Although details of the non-lytic exocytosis phenomenon in amoeba and macrophages differ, such as their discordant response to cytochalasin D, the finding of same general phenomenon in such evolutionarily distant cells, one of which is commonly found in cryptococcal ecologic niches, suggests that this mechanism was initially selected for escape from environmental predators.

Other Single Celled Hosts for Fungi

In addition to amoeba two other single celled hosts have been evaluated for their interactions with *C. neoformans*: *Paramecium* spp. (Frager et al. 2010) and *Dictyostelium discoideum* (Steenbergen et al. 2003). Like some species of amoebae, *Paramecium* spp. were efficient grazers of *C. neoformans*, ingesting and digesting fungal cells. Hence, apart from a temporary cessation of movement after Paramecia encountered the fungal cells, no damage was apparent from their interaction with *C. neoformans*. Like *A. castellani*, *D. discoideum* ameboid cells were susceptible to *C. neoformans* although the much smaller cells of this social amoeba were significantly less efficient at phagocytosing yeast cells. Co-cultivation of an avirulent *C. neoformans* cells with *D. discoideum* cells resulted in an increase in virulence for mice, implying that exposure to *C. neoformans* to social amoebae cells resulted in changes that translated into increased fitness in the mammalian host. Although the mechanism responsible for the increase in virulence is not understood the relatively short incubation interval suggests that this effect did not involve selection for more virulent genetic variants but was perhaps mediated by epigenetic changes.

The studies with *D. discoidum* revealed that the concept of microbial opportunism could be extended to unicellular level. *C. neoformans* acapsular mutants that were not pathogenic for wild type *D. discoideum* were pathogenic for slime mold strains defective myosin VII suggesting that notion that avirulent microbes for normal hosts can be virulent for impaired hosts also occurs at the unicellular level (Steenbergen et al. 2003). The experience with *Paramecium* spp. and *D. discoiduem* suggests that the outcome of the protist model host interaction with *C. neoformans* is likely to reflect the specific characteristics of the interacting pair and it is likely that each different fungal–protist interaction is unique. This concept is important because there are innumerable protist species in the environment and caution is warranted in generalizing results from specific interactions.

Thoughts Towards a Synthesis on the Origin of Virulence

The observations that: (1) many aspects of *C. neoformans* interactions with amoebae closely resemble interactions with macrophages (Steenbergen et al. 2001; Zaragoza et al. 2008); (2) virulence factors of *C. neoformans* for mammals are also virulence factors for amoebae (Steenbergen et al. 2001);

(3) complex mechanisms such as non-lytic exocytosis from macrophages has a correlate in amoebae (Chrisman et al. 2010); (4) virulence of *C. neoformans* and *H. capsulatum* for mice can be enhanced by passage in *D. discoideum* (Steenbergen et al. 2004) and *A. castellani* (Steenbergen et al. 2004), respectively; and (5) yeast-hyphal transitions in dimorphic fungi can be triggered at non-permissive temperatures (Steenbergen et al. 2004), support the hypothesis that the capacity for virulence in these microbes was selected and maintained by selection pressures in the environment, including amoebae predation. This view has been proposed and expanded upon on several publications and opinion articles (Steenbergen et al. 2001, 2003; Casadevall et al. 2003; Steenbergen and Casadevall 2003; Casadevall and Pirofski 2007). A similar view has been proposed to account for the anti-phagocytic behavior of entomogenous fungi that prey on insects (Bidochka et al. 2010). However, consideration of this hypothesis immediately leads to another question: if all soil microbes are under similar selection pressures that can lead to the acquisition of trails suitable for animal virulence, how come there are so few pathogens relative to the total number of microbial species in soils? In considering possible answers it is noteworthy that different microbial species by definition are constituted of microbes with different genetic, biochemical and cellular characteristics and that only a subset of these traits is capable of interacting with an animal host in a manner where the outcome is sufficient damage to the host to manifest itself as disease. In this regard, I have proposed that soil microbes could be considered to playing a metaphorical card game where each holds a different 'hand' of attributes (Casadevall 2006). Some card 'hands' translate into the capacity for animal, plant, or insect virulence, but the majority of soil microbes are holding on to 'hands' that allow their survival but does not confer the capacity for virulence (Casadevall 2006). According to the view the capacity for virulence may emerge stochastically, by the acquisition of traits that enable environmental survival in a process that can also be referred as accidental virulence (Casadevall and Pirofski 2007).

References

Allen PG, Dawidowicz EA (1990a) Phagocytosis in Acanthamoeba: I. A mannose receptor is responsible for the binding and phagocytosis of yeast. J Cell Physiol 145:508–513

Allen PG, Dawidowicz EA (1990b) Phagocytosis in Acanthamoeba: II. Soluble and insoluble mannose-rich ligands stimulate phosphoinositide metabolism. J Cell Physiol 145:514–521

Alvarez M, Casadevall A (2006) Phagosome fusion and extrusion, and host cell survival following *Cryptococcus neoformans* phagocytosis by macrophages. Curr Biol 16:2161–2165

Alvarez M, Burn T, Luo Y, Pirofski LA, Casadevall A (2009) The outcome of *Cryptococcus neoformans* intracellular pathogenesis in human monocytes. BMC Microbiol 9:51

Bergman A, Casadevall A (2010) Mammalian endothermy optimally restricts fungi and metabolic costs. MBio 1:e00212-10

Bidochka MJ, Clark DC, Lewis MW, Keyhani NO (2010) Could insect phagocytic avoidance by entomogenous fungi have evolved via selection against soil amoeboid predators? Microbiology 156:2164–2171

Blehert DS et al (2009) Bat white-nose syndrome: an emerging fungal pathogen? Science 323:227

Bowers B, Olszewski TE (1983) *Acanthamoeba* discriminates internally between digestible and indigestible particles. J Cell Biol 97:317–322

Bunting LA, Neilson JB, Bulmer GS (1979) *Cryptococcus neoformans*: gastronomic delight of a soil ameba. Sabouraudia 17:225–232

Casadevall A (2006) Cards of virulence and the global virulome for humans. Microbe 1:359–364

Casadevall A (2007) The cards of virulence and the global virulome. Microbe 1:359–364

Casadevall A, Pirofski L (1999) Host-pathogen interactions: redefining the basic concepts of virulence and pathogenicity. Infect Immun 67:3703–3713

Casadevall A, Pirofski L (2000) Host-pathogen interactions: the basic concepts of microbial commensalism, colonization, infection, and disease. Infect Immun 68:6511–6518

Casadevall A, Pirofski L (2001) Host-pathogen interactions: the attributes of virulence. J Infect Dis 184:337–344

Casadevall A, Pirofski L (2003) The damage-response framework of microbial pathogenesis. Nat Microbiol Rev 1:17–24

Casadevall A, Pirofski LA (2007) Accidental virulence, cryptic pathogenesis, martians, lost hosts, and the pathogenicity of environmental microbes. Eukaryot Cell 6:2169–2174

Casadevall A, Nosanchuk JD, Steenbergen JN (2003) 'Ready-made' virulence and 'dual-use' virulence factors in pathogenic enviromental fungi - the *Cryptococcus neoformans* paradigm. Curr Opin Microbiol 112:1164–1175

Castellani A (1931) An amoeba growing in cultures of a yeast. J Trop Med Hyg 33:188–191

Castellani A (1955) Phagocytic and destructive action of *Hartmanella castellanii* (Amoeba castellanii) on pathogenic encapsulated yeast-like fungus *Torulopsis neoformans* (*Cryptococcus neoformans*). Ann Inst Pasteur 89:1–7

Chaturvedi V et al (2010) Morphological and molecular characterizations of psychrophilic fungus *Geomyces destructans* from New York bats with White Nose Syndrome (WNS). PLoS One 5:e10783

Chrisman CJ, Alvarez M, Casadevall A (2010) Phagocytosis and non-lytic phagocytosis of *Cryptococcus neoformans* by, and from, Acanthamoeba castellanii. Appl Environ Microbiol 76:6056–6062

Cleaveland S, Laurenson MK, Taylor LH (2001) Diseases of humans and their domestic mammals: pathogen characteristics, host range and the risk of emergence. Philos Trans R Soc Lond B Biol Sci 356:991–999

Couteaux M-M, Darbyshire JF (1988) Functional diversity among soil protozoa. Appl Soil Ecol 10:229–237

Currie BP, Freundlich LF, Casadevall A (1994) Restriction fragment length polymorphism analysis of *Cryptococcus neoformans* isolates from environmental (pigeon excreta) and clinical isolates in New York City. J Clin Microbiol 32:1188–1192

Ettinger MR et al (2003) Distribution of free-living amoebae in James River, Virginia, USA. Parasitol Res 89:6–15

Fan W, Idnurm A, Breger J, Mylonakis E, Heitman J (2007) Eca1, a sarcoplasmic/endoplasmic reticulum Ca^{2+}–ATPase, is involved in stress tolerance and virulence in *Cryptococcus neoformans*. Infect Immun 75:3394–3405

Feldmesser M, Kress Y, Novikoff P, Casadevall A (2000) *Cryptococcus neoformans* is a facultative intracellular pathogen in murine pulmonary infection. Infect Immun 68:4225–4237

Feldmesser M, Tucker SC, Casadevall A (2001) Intracellular parasitism of macrophages by *Cryptococcus neoformans*. Trends Microbiol 9:273–278

Frager SZ, Chrisman CJ, Shakked R, Casadevall A (2010) Paramecium species ingest and kill the cells of the human pathogenic fungus Cryptococcus neoformans. Med Mycol 48:775–779

Garcia-Solache MA, Casadevall A (2010) Global warming will bring new fungal diseases for mammals. MBio 1:e00061-10

Goldman DL et al (2001) Serologic evidence for *Cryptococcus infection* in early childhood. Pediatrics 107:E66

Ma H, Croudace JE, Lammas DA, May RC (2006) Expulsion of live pathogenic yeast by macrophages. Curr Biol 16:2156–2160

Malliaris SD, Steenbergen JN, Casadevall A (2004) Cryptococcus neoformans var. gattii can exploit Acanthamoeba castellanii for growth. Med Mycol 42:149–158

Napolitano JJ (1982) Isolation of amoebae from edible mushrooms. Appl Environ Microbiol 44:255–257

Neilson JB, Ivey MH, Bulmer GS (1978) *Cryptococcus neoformans*: pseudohyphal forms surviving culture with *Acanthamoeba polyphaga*. Infect Immun 20:262–266

Nero LC, Tarver MG, Hedrick LR (1964) Growth of *Acanthomoeba castellani* with the yeast *Torulopsis famata*. J Bacteriol 87:220–225

Nielsen K et al (2005) Cryptococcus neoformans alpha strains preferentially disseminate to the central nervous system during coinfection. Infect Immun 73:4922–4933

Perfect JR, Lang SDR, Durack DT (1980) Chronic cryptococcal meningitis. Am J Pathol 101:177–193

Rigothier MC, Maccario J, Gayral P (1994) Inhibitory activity of saccharomyces yeasts on the adhesion of Entamoeba histolytica trophozoites to human erythrocytes in vitro. Parasitol Res 80:10–15

Rivera F et al (1984) Pathogenic and free-living protozoa cultured from the nasopharyngeal and oral regions of dental patients. Environ Res 33:428–440

Robert VA, Casadevall A (2009) Vertebrate endothermy restricts most fungi as potential pathogens. J Infect Dis 200:1623–1626

Robinson BS, Bamforth SS, Dobson PJ (2002) Density and diversity of protozoa in some arid Australian soils. J Eukaryot Microbiol 49:449–453

Ruiz A, Neilson JB, Bulmer GS (1982) Control of *Cryptococcus neoformans* in nature by biotic factors. Sabouraudia 20:21–29

Sheehan KB, Fagg JA, Ferris MJ, Henson JM (2003) PCR detection and analysis of the free-living amoeba Naegleria in hot springs in Yellowstone and Grand Teton National Parks. Appl Environ Microbiol 69:5914–5918

Steenbergen JN, Casadevall A (2003) The origin and maintenance of virulence for the human pathogenic fungus *Cryptococcus neoformans*. Microbes Infect 5:667–675

Steenbergen JN, Shuman HA, Casadevall A (2001) *Cryptococcus neoformans* interactions with amoebae suggest an explanation for its virulence and intracellular pathogenic strategy in macrophages. Proc Natl Acad Sci USA 18:15245–15250

Steenbergen JN, Nosanchuk JD, Malliaris SD, Casadevall A (2003) *Cryptococcus neoformans* virulence is enhanced after intracellular growth in the genetically malleable host *Dictyostelium discoideum*. Infect Immun 71:4862–4872

Steenbergen JN, Nosanchuk JD, Malliaris SD, Casadevall A (2004) Interaction of *Blastomyces dermatitidis*, *Sporothrix schenckii*, and *Histoplasma capsulatum* with *Acanthamoeba castellanii*. Infect Immun 72:3478–3488

Taylor LH, Latham SM, Woolhouse ME (2001) Risk factors for human disease emergence. Philos Trans R Soc Lond B Biol Sci 356:983–989

Upadhyay JM (1968) Growth and bacteriolytic activity of a soil amoeba, Hartmannella glebae. J Bacteriol 95:771–774

Zaragoza O et al (2008) Capsule enlargement in *Cryptococcus neoformans* confers resistance to oxidative stress suggesting a mechanism for intracellular survival. Cell Microbiol 10:2043–2057

Chapter 2
Of Model Hosts and Man: Using *Caenorhabditis elegans*, *Drosophila melanogaster* and *Galleria mellonella* as Model Hosts for Infectious Disease Research

Justin Glavis-Bloom, Maged Muhammed, and Eleftherios Mylonakis

Abstract The use of invertebrate model hosts has increased in popularity due to numerous advantages of invertebrates over mammalian models, including ethical, logistical and budgetary features. This review provides an introduction to three model hosts, the nematode *Caenorhabditis elegans*, the fruit fly *Drosophila melanogaster* and the larvae of *Galleria mellonella*, the greater wax moth. It highlights principal experimental advantages of each model, for *C. elegans* the ability to run high-throughput assays, for *D. melanogaster* the evolutionarily conserved innate immune response, and for *G. mellonella* the ability to conduct experiments at 37°C and easily inoculate a precise quantity of pathogen. It additionally discusses recent research that has been conducted with each host to identify pathogen virulence factors, study the immune response, and evaluate potential antimicrobial compounds, focusing principally on fungal pathogens.

Introduction

The study of infectious disease requires model hosts. For obvious ethical reasons it is impossible to conduct human in-vivo primary experimentation to identify pathogen virulence factors, study the immune response to pathogenic infection, or evaluate potential antimicrobial compounds for toxicity and effectiveness. The murine model *Mus musculus* has long been a favored model host, as it provides a similarity of human and mouse anatomy, immune response, and in some cases pathogen susceptibility. Yet there are many drawbacks to the mouse model. First, there are ethical concerns with mammalian experimentation. Second, there are logistical obstacles, including lengthy reproduction time and the difficulty and expense associated with obtaining and maintaining sufficient numbers of mice to conduct experimentation.

Fortunately, evolutionary conservation extends from humans to distantly related metazoans, permitting the use of invertebrates as model hosts for pathogenesis studies. Invertebrates have numerous advantages over mammalian models. Invertebrates can be inexpensively obtained, easily

J. Glavis-Bloom • M. Muhammed
Division of Infectious Diseases, Massachusetts General Hospital, Boston, MA 02114, USA

E. Mylonakis (✉)
Harvard Medical School, Division of Infectious Disease, Massachusetts General Hospital, 55 Fruit street, Gray-Jackson 5, Room GRJ-504, Boston, MA 02114, USA
e-mail: emylonakis@partners.org

E. Mylonakis et al. (eds.), *Recent Advances on Model Hosts*, Advances in Experimental Medicine and Biology 710, DOI 10.1007/978-1-4419-5638-5_2, maintained, and their small size and short lifespan facilitates experimentation in the laboratory setting. For these reasons, the use of invertebrate model hosts has been increasing in popularity. This review covers three invertebrate model hosts, describing principal experimental advantages and highlighting recent research. The soil-living nematode *Caenorhabditis elegans* is an ideal model for high-throughput screening due to its small size, transparent body, rapid reproductive cycle producing genetically identical progeny, and fully sequenced genome with available genomic tools. The fruit fly *Drosophila melanogaster* mounts a sophisticated innate immune response that is homologous to the mammalian innate immune response and, like *C. elegans*, has a fully sequenced genome and available genomic tools. Lastly, the larvae of *Galleria mellonella*, the greater wax moth, can survive at 37°C (thus providing an analog to humans when studying temperature-sensitive pathogen virulence) and, due to its relatively large size, can easily be inoculated with a precise quantity of pathogen using a syringe.

Caenorhabditis elegans

In the 1960s, Sydney Brenner established the soil-living nematode *Caenorhabditis elegans* as a genetic model host with tremendous potential for studying cell biology and genetics in-vivo (Brenner 1974). Its small size (adults are ~1 mm), rapid life cycle (~3.5 days at 20°C), and transparent body make the organism well suited to experimental observation. Other important features of *C. elegans* include its ability to produce genetically identical progeny as a self-fertilizing hermaphrodite, its small and fully sequenced genome, and its physiological and anatomical simplicity (~1,000 fully-mapped cells, including ~300 neurons) (Riddle et al. 1997). In addition, as discovered by Fire and Mello, delivering double-stranded RNA-mediated interference (RNAi) by feeding is a means of genetic disruption in *C. elegans* (Fire et al. 1998). Also, *C. elegans* can be maintained indefinitely in liquid nitrogen, a fact that led to the creation of libraries of thousands of easily and inexpensively obtainable mutant strains (Bazopoulou and Tavernarakis 2009). Most importantly, *C. elegans* is uniquely well suited to study infectious agents because it naturally feeds on microorganisms and because it is susceptible to many of the same bacterial and fungal pathogens that can kill mammals and humans (O'Callaghan and Vergunst 2010). In laboratory, *C. elegans* is usually fed a lawn of non-pathogenic *Escherichia coli*, which can be substituted with pathogenic bacteria or fungi to generate an infection and test genetic virulence factors or screen potential antimicrobial compounds. Broadly, in-vivo screening in the microbiology field using *C. elegans* can be utilized in three ways to: (1) Identify the genetic basis of pathogen virulence; (2) Gain insight into the immune response; and (3) Identify potential antimicrobial compounds.

One method of discovering the genetic basis of bacterial virulence is screening libraries of various bacterial mutants to identify those with increased or decreased virulence. The gram-negative bacterium *Pseudomonas aeruginosa* is a free-living opportunistic pathogen capable of causing mortality in immunocompromised patients, and a leading cause of hospital-acquired and ventilator-associated pneumonia, with a high degree of intrinsic virulence and multi-drug resistance (Diaz et al. 2008). *P. aeruginosa* transposon mutation libraries have been screened for mutant clones that exhibit a reduced ability to kill *C. elegans*, and many of the mutants identified in this way are also less virulent in murine models of infection (Mahajan-Miklos et al. 1999; Tan et al. 1999). In a separate study, a high throughput screen of 2,200 *P. aeruginosa* mutants using a liquid infection assay showed attenuated virulence associated with a mutation in the *cheB2* gene. This finding was subsequently confirmed in a murine lung infection model, illustrating the applicability of bacterial virulence findings in *C. elegans* to mammals (Garvis et al. 2009). Similar work has been done with *Serratia marcescens*, another gram-negative bacterium frequently associated with hospital-acquired urinary tract infections, and with *Staphylococcus aureus*, a gram-positive bacterium that has increasing antibiotic resistance, to identify the genetic mutations associated with attenuated virulence in *C. elegans* (Kurz et al. 2003; Begun et al. 2005).

Furthermore, several studies have been done with *C. elegans* assays to identify virulence factors associated with two of the most prevalent fungal pathogens, *Candida albicans* and *Cryptococcus neoformans*. *C. albicans* is an opportunistic fungal pathogen commonly carried in the human gastrointestinal tract with harmful effects in immunocompromised patients. It is the fourth most common cause of bloodstream infection, with costly treatment and high mortality rates (35%) (Douglas 2003). *C. albicans* has the ability to undergo morphological change from a yeast form to a hyphal form. Using a *C. elegans* assay, mutant *C. albicans* strains with diminished hyphal formation capability were found to have attenuated virulence, and several genes were identified that are important for hyphal formation in vivo (Pukkila-Worley et al. 2009). *C. neoformans* is the third most common cause of invasive fungal infections in solid organ transplant recipients and can be life threatening in immunocompromised patients (Mueller and Fishman 2003). Using a *C. elegans* screen, 350 *C. neoformans* mutants were tested for reduced virulence. Among seven mutants identified, one contained an insertion in a gene encoding a serine/threonine protein kinase (KIN1), which is also an important virulence factor in murine models of infection (Mylonakis et al. 2004).

Increasing microbial resistance to many antibiotics and antifungal agents has created a need to identify new compounds for therapeutic use. *C. elegans* provides an ideal screening model to identify potential antimicrobial candidates in vivo. Because of its small size *C. elegans* is particularly well-suited to high throughput screening in standard 348 well plates of thousands of compounds to identify those with potential novel antimicrobial activities (Spring 2005). In vitro high throughput screens against specific microbial targets have produced high attrition rates due to an inability to forecast preclinical and clinical development barriers, including toxicity of the compound under the study (Lindsay 2003). In addition, traditional antibiotic screens for compounds that block pathogen growth in vitro cannot identify compounds that stimulate immune responses or decrease pathogen virulence during infection (often referred to immunomodulatory and antivirulence compounds, respectively). In contrast, in vivo whole animal screens for compounds that cure an infection immediately identify toxic compounds and other potential clinical obstacles, and can identify compounds that may enhance host immunity (Moy et al. 2009). The small size and transparent body of *C. elegans* also facilitates the use of robots to dispense a precise number of live, *C. elegans* into microwells and facilitates automatic image analysis to identify the number of surviving organisms during the experiment (Moy et al. 2009).

An automated *C. elegans* assay has been used to identify compounds that cure infections caused by the bacterium *Enterococcus faecalis* or the yeast *Candida albicans*. First, in a screen of 372,000 compounds with *E. faecalis*, a gram-positive bacterium that is increasingly acquiring resistance to antibiotics, 28 compounds were identified that were not previously reported to have antimicrobial properties, including at least 6 which affected the growth of the pathogen in vivo, but not in vitro, suggesting they act by mechanisms distinct from antibiotics currently in use (Moy et al. 2009). In the case of *C. albicans*, a screen of 2,560 natural compounds resulted in the identification of 12 saponins, which significantly enhanced survival of the worms, suggesting that saponins have the potential to form a foundation for a new generation of antifungal compounds (Coleman et al. 2010). Another screen of 3,228 bioactive compounds yielded 19 compounds that resulted in an increase in *C. elegans* survival after infection with *C. albicans*. Of these, 12 were not primarily used as antifungal agents, including 3 immunosuppressive drugs (Okoli et al. 2009). An earlier screen of 1,266 compounds for antifungal activity identified 15 compounds that prolonged survival of *C. elegans* after infection with *C. albicans* and inhibited hyphal formation. Two of these compounds, caffeic acid phenethyl ester, a major active component of honeybee propolis, and the fluoroquinolone enoxacin, also attenuated *C. albicans* virulence in a murine model (Breger et al. 2007).

Although *C. elegans* has numerous advantages, the organism does pose some limitations as a model host. In the case of compound screens, it can be difficult to predict mammalian bioactivity given the anatomical simplicity of *C. elegans* relative to mammals, and challenging to forecast effective concentrations of identified compounds, as the nematode's thick cuticle blocks absorption and its small size makes it impossible to measure the concentration of compound that has been absorbed (Giacomotto and Ségalat 2010). Additionally, some diseases and immune responses cannot be recreated because the nematode lacks a variety of mammalian anatomical structures. To circumvent these limitations, other models must be used.

Drosophila melanogaster

The fruit fly *Drosophila melanogaster* has many of the same advantages as *C. elegans*, including small size, short generation time, a fully sequenced genome, and pre-existing libraries of genetic mutants. In addition, *D. melanogaster* is an excellent model host because it mounts an extensively studied innate immune response, with genes and pathways similar to those found in mammals (Hoffmann 2003). In particular, the Toll and Imd (immune deficiency) pathways in *D. melanogaster* are useful models for mammalian study, given the similarity of the Toll receptor to mammalian Toll-like receptors (TLR) and interleukin-1 (IL-1) receptors, and the similarity of the Imd pathway to the mammalian tumor necrosis factor signaling pathway (Sekiya et al. 2008).

The innate immune response in *D. melanogaster* is comprised of both cellular and humoral components. The cellular response involves specialized hemocytes (blood cells), which engage in phagocytosis and encapsulation of foreign microbes (Rizki and Rizki 1984). The humoral response involves the production of antimicrobial peptides (AMPs) in the fat body, the equivalent of the mammalian liver, which are then secreted into the haemolymph (Lemaitre 2004). Approximately 20 AMPs that have been discovered can be classified into seven groups, with differential effectiveness against fungi (Drosomycin and Metchnikowin), gram-positive bacteria (Defensin), and gram-negative bacteria (Diptericin, Drosocin, Attacin and Cecropin) (Lemaitre and Hoffmann 2007). The regulation of the genes encoding AMP production occurs via the Toll and Imd signaling pathways. The Toll pathway is activated primarily in response to fungal and some gram-positive bacterial infections, whereas the Imd pathway is activated predominantly in response to gram-negative and some gram-positive bacterial infections (Lemaitre et al. 1997).

Utilizing *D. melanogaster* in studies with fungal pathogens, including *C. albicans*, *Cryptococcus neoformans* and *Aspergillus fumigatus,* has provided insights into the innate immune response and mechanisms of pathogenic virulence (Chamilos et al. 2006). These studies have also aided in establishing *D. melanogaster* as a potential model for testing antimicrobial compounds efficacy. *D. melanogaster* mutants devoid of functioning Toll receptors are analogous in many ways to immunocompromised mammals that are at risk for infection by opportunistic fungi. For example, a study of *C. albicans* mutant strains revealed the same rank order of virulence in Toll-deficient *D. melanogaster* as in mammalian infection models, and moreover showed that increased virulence in *Drosophila* was associated with the ability of the mutant *C. albicans* strain to form hyphae, the same virulence mechanism as is demonstrated in mammals (Alarco et al. 2004). Similarly, *D. melanogaster* has proven to be an important model host in studies of the pathogenic fungi *C. neoformans*. In mammals, a significant aspect of *C. neoformans* virulence is its ability to avoid phagocytosis and the immune response via polysaccharide capsule formation. In a study of mutant strains of *C. neoformans* in *D. melanogaster*, genes associated with pathogenesis in mammals caused enhanced killing of *D. melanogaster*, but *C. neoformans* with a mutated gene essential for capsule formation (cap59) exhibited only a minor decrease in virulence, suggesting factors other than capsular formation contribute to *C. neoformans* virulence in *D. melanogaster* (Apidianakis et al. 2004). Additional research has established *D. melanogaster* as a potential model host for in vivo assessment of antifungal compound efficacy. Although *D. melanogaster* cannot be grown in liquid medium, which precludes its use in robotic high throughput screening, screening of potential compounds can be done by hand (Giacomotto and Ségalat 2010). Two studies in particular evaluated the response of *D. melanogaster* infected with *A. fumigatus* and *C. albicans*, finding *D. melanogaster* a reliable model of testing antifungal compounds currently in use, and a potential model for testing combinations of antifungal drugs (Lionakis et al. 2005; Chamilos et al. 2006).

Galleria mellonella

The larval stage of the greater wax moth *Galleria mellonella* presents unique advantages as a model host. Chief among these advantages is its ability to survive at 37°C, thus providing an analog to humans when studying pathogenic temperature-sensitive virulence. Changes in temperature, however, can provoke the *G. mellonella* immune response (Mowlds and Kavanagh 2008). *G. mellonella* caterpillars can be stored at room temperature in the lab, and are easily and inexpensively obtained in sizes large enough (1.5–2.5 cm) to be inoculated by hand with a syringe, permitting the delivery of a precise amount of pathogen (Kavanagh and Fallon 2010). Significant limitations of the G. *mellonella* model are that its genome has not yet been sequenced and well-established methods of generating mutants have not been developed. Yet, unlike wild-type *D. melanogaster*, which can survive large inocula of pathogen (often necessitating the use of Toll mutants to study pathogenesis), *G. mellonella* is susceptible to infection by numerous pathogens (Mylonakis 2008). In particular, *G. mellonella* serves as an ideal model host for studying pathogen virulence mechanisms and the efficacy of potential antimicrobial compounds.

G. mellonella larvae can be killed by infection with *Candida*, and the hierarchy of virulence of different *Candida* species in *G. mellonella* is consistent with virulence observed in mammalian models, with *C. albicans* demonstrating the greatest pathogenicity (Cotter et al. 2000). *C. albicans* virulence is associated with the ability to form hyphae, and a good correlation exists between *C. albicans* mutants with decreased hyphal formation and decreased virulence in both *G. mellonella* and mice (Brennan et al. 2002). Recent work with *G. mellonella*, however, has demonstrated that hyphal formation alone is insufficient to kill *G. mellonella*, as mutant *C. albicans* strains exist with the ability to form filaments with impaired virulence (Fuchs et al. 2010).

G. mellonella larvae can also be employed to evaluate the effectiveness of antifungal compounds and treatment protocols. For example, following infection with *C. neoformans*, *G. mellonella* was inoculated in one study with amphotericin B, flucytosine, and fluconazole. Treatment guidelines for severe cryptococcal infection in humans call for the combination of amphotericin B plus flucytosine, which was also associated with greatest *G. mellonella* survival, suggesting *G. mellonella* may be a valuable model host for future *C. neoformans* antifungal compound discovery (Mylonakis et al. 2005). Similarly, *G. mellonella* larvae were used to assess the efficacy of silver (I) and 1,10-phenanthroline after infection with *C. albicans*, and demonstrated significantly increased survival with both compounds (Rowan et al. 2009). Finally, an infection assay in *G. mellonella* with *C. albicans* and *A. fumigatus* revealed that co-inoculation of an Hsp90 inhibitor enhanced the efficacy of existing antifungal drugs (Cowen et al. 2009).

Conclusion

It is becoming increasingly critical to understand pathogenic virulence factors and identify novel therapeutic options as the population of immunocompromised patients grows and new pathogenic resistance to conventional antimicrobial therapies emerges. Invertebrate model hosts will continue to play important roles in the development of new lifesaving treatments and in finding virulence traits that could be potential targets for new antimicrobials. In particular, further robotic-assisted high throughput assays with *C. elegans* will be critical in identifying new antimicrobial compounds, our understanding of the innate immune system will increase with further experimentation with *D. melanogaster* and, as *G. mellonella* becomes a more established model, genomic tools will emerge that will further increase its usefulness. In addition, new model hosts will continue to be discovered that will further our understanding and assist in the development of novel lifesaving therapies.

References

Alarco AM et al (2004) Immune-deficient *Drosophila melanogaster*: a model for the innate immune response to human fungal pathogens. J Immunol 172:5622–5628

Apidianakis Y et al (2004) Challenge of *Drosophila melanogaster* with *Cryptococcus neoformans* and role of the innate immune response. Eukaryot Cell 3:413–419

Bazopoulou D, Tavernarakis N (2009) The NemaGENETAG initiative: large scale transposon insertion gene-tagging in *Caenorhabditis elegans*. Genetica 137:39–46

Begun J, Sifri CD, Goldman S, Calderwood SB, Ausubel FM (2005) *Staphylococcus aureus* virulence factors identified by using a high-throughput *Caenorhabditis elegans*-killing model. Infect Immun 73:872–877

Breger J et al (2007) Antifungal chemical compounds identified using a *C. elegans* pathogenicity assay. PLoS Pathog 3:e18

Brennan M, Thomas DY, Whiteway M, Kavanagh K (2002) Correlation between virulence of *Candida albicans* mutants in mice and *Galleria mellonella* larvae. FEMS Immunol Med Microbiol 34:153–157

Brenner S (1974) The genetics of *Caenorhabditis elegans*. Genetics 77:71–94

Chamilos G et al (2006) *Drosophila melanogaster* as a facile model for large-scale studies of virulence mechanisms and antifungal drug efficacy in *Candida* species. J Infect Dis 193:1014–1022

Coleman JJ et al (2010) Characterization of plant-derived saponin natural products against *Candida albicans*. ACS Chem Biol 5:321–332

Cotter G, Doyle S, Kavanagh K (2000) Development of an insect model for the in vivo pathogenicity testing of yeasts. FEMS Immunol Med Microbiol 27:163–169

Cowen LE et al (2009) Harnessing Hsp90 function as a powerful, broadly effective therapeutic strategy for fungal infectious disease. Proc Natl Acad Sci USA 106:2818–2823

Diaz MH et al (2008) *Pseudomonas aeruginosa* induces localized immunosuppression during pneumonia. Infect Immun 76:4414–4421

Douglas LJ (2003) *Candida* biofilms and their role in infection. Trends Microbiol 11:30–36

Fire A et al (1998) Potent and specific genetic interference by double-stranded RNA in *Caenorhabditis elegans*. Nature 391:806–811

Fuchs BB et al (2010) Role of filamentation in *Galleria mellonella* killing by *Candida albicans*. Microbes Infect 12:488–496

Garvis S et al (2009) *Caenorhabditis elegans* semi-automated liquid screen reveals a specialized role for the chemotaxis gene cheB2 in *Pseudomonas aeruginosa* virulence. PLoS Pathog 5:e1000540

Giacomotto J, Ségalat L (2010) High-throughput screening and small animal models, where are we? Br J Pharmacol 160:204–216

Hoffmann J (2003) The immune response of *Drosophila*. Nature 426:33–38

Kavanagh K, Fallon JP (2010) *Galleria mellonella* larvae as models for studying fungal virulence. Fungal Biol Rev 24:79–83

Kurz CL et al (2003) Virulence factors of the human opportunistic pathogen *Serratia marcescens* identified by *in vivo* screening. EMBO J 22:1451–1460

Lemaitre B (2004) The road to toll. Nat Rev Immunol 4:521–527

Lemaitre B, Hoffmann J (2007) The host defense of *Drosophila melanogaster*. Annu Rev Immunol 25:697–743

Lemaitre B, Reichhart JM, Hoffmann JA (1997) *Drosophila* host defense: differential induction of antimicrobial peptide genes after infection by various classes of microorganisms. Proc Natl Acad Sci USA 23:14614–14619

Lindsay MA (2003) Target discovery. Nat Rev Drug Discov 2:831–838

Lionakis MS et al (2005) Toll-deficient *Drosophila* flies as a fast, high-throughput model for the study of antifungal drug efficacy against invasive aspergillosis and *Aspergillus* virulence. J Infect Dis 191:1188–1195

Mahajan-Miklos S, Tan MW, Rahme LG, Ausubel FM (1999) Molecular mechanisms of bacterial virulence elucidated using a *Pseudomonas aeruginosa-Caenorhabditis elegans* pathogenesis model. Cell 96:47–56

Mowlds P, Kavanagh K (2008) Effect of pre-incubation temperature on susceptibility of *Galleria mellonella* larvae to infection by *Candida albicans*. Mycopathologia 165:5–12

Moy T et al (2009) High-throughput screen for novel antimicrobials using a whole animal infection model. ACS Chem Biol 4:527–533

Mueller NJ, Fishman JA (2003) Asymptomatic pulmonary cryptococcosis in solid organ transplantation: report of four cases and review of the literature. Transpl Infect Dis 5:140–143

Mylonakis E (2008) *Galleria mellonella* and the study of fungal pathogenesis: making the case for another genetically tractable model host. Mycopathologia 165:1–3

Mylonakis E et al (2004) *Cryptococcus neoformans* Kin1 protein kinase homologue, identified through a *Caenorhabditis elegans* screen, promotes virulence in mammals. Mol Microbiol 54:407–419

Mylonakis E et al (2005) *Galleria mellonella* as a model system to study *Cryptococcus neoformans* pathogenesis. Infect Immun 73:3842–3850

O'Callaghan D, Vergunst A (2010) Non-mammalian animal models to study infectious disease: worms or fly fishing? Curr Opin Microbiol 13:79–85

Okoli I et al (2009) Identification of antifungal compounds active against *Candida albicans* using an improved high-throughput *Caenorhabditis elegans* assay. PLoS One 4:e7025

Pukkila-Worley R, Peleg A, Tampakakis E, Mylonakis E (2009) *Candida albicans* hyphal formation and virulence assessed using a *Caenorhabditis elegans* infection model. Eukaryot Cell 8:1750–1758

Riddle DL, Blumenthal T, Meyer BG, Priess JR (1997) *C. elegans II*, 2nd edn. Cold Spring Harbor Laboratory Press, Plainview

Rizki RM, Rizki TM (1984) Selective destruction of a host blood cell type by a parasitoid wasp. Proc Natl Acad Sci USA 81:6154–6158

Rowan R, Moran C, McCann M, Kavanagh K (2009) Use of *Galleria mellonella* larvae to evaluate the in vivo antifungal activity of [Ag2(mal)(phen)3]. Biometals 22:461–467

Sekiya M et al (2008) A cyclopentanediol analogue selectively suppresses the conserved innate immunity pathways, *Drosophila* IMD and TNF-alpha pathways. Biochem Pharmacol 75:2165–2174

Spring DR (2005) Chemical genetics to chemical genomics: small molecules offer big insights. Chem Soc Rev 34:472–482

Tan MW, Rahme LG, Sternberg JA, Tompkins RG, Ausubel FM (1999) *Pseudomonas aeruginosa* killing of *Caenorhabditis elegans* used to identify *P. aeruginosa* virulence factors. Proc Natl Acad Sci USA 96:2408–2413

Chapter 3
Caenorhabditis elegans as an Alternative Model to Study Senescence of Host Defense and the Prevention by Immunonutrition

Tomomi Komura, Takanori Ikeda, Kaori Hoshino, Ayumi Shibamura, and Yoshikazu Nishikawa

Abstract Whether nutritional control can retard senescence of immune function and decrease mortality from infectious diseases has not yet been established; the difficulty of establishing a model has made this a challenging topic to investigate. *Caenorhabditis elegans* has been extensively used as an experimental system for biological studies. Particularly for aging studies, the worm has the advantage of a short and reproducible life span. The organism has also been recognized as an alternative to mammalian models of infection with bacterial pathogens in this decade. Hence we have studied whether the worms could be a model host in the fields of immunosenescence and immunonutrition. Feeding nematodes lactic acid bacteria (LAB) resulted in increases in average life span of the nematodes compared to those fed *Escherichia coli* strain OP50, a standard food bacteria. The 7-day-old nematodes fed LAN from age 3 days were clearly endurable to subsequent salmonella infection compared with nematodes fed OP50 before the salmonella infection. The worm could be a unique model to study effects of food factors on longevity and host defense, so-called immunonutrition. Then we attempted to establish an immunosenescence model using *C. elegans*. We focused on the effects of worm age on the *Legionella* infection and the prevention by immunonutrition. No significant differences in survival were seen between 3-day-old worms fed OP50 and 3-day-old worms infected with virulent *Legionella* strains. However, when the worms were infected from 7.5 days after hatching, the virulent *Legionella* strains were obviously nematocidal for the worms' immunosenescence. In contrast, nematodes fed with bifidobacteria prior to *Legionella* infection were resistant to *Legionella*. *C. elegans* could act as a unique alternative host for immunosenescence and resultant opportunistic infection, and immunonutrition researches.

Introduction

Caenorhabditis elegans is a small free-living soil nematode that feeds on bacteria; it has been extensively used as an experimental system for biological studies because of its simplicity, transparency, ease of cultivation, and suitability for genetic analysis (Riddle et al. 1997). Particularly for aging studies,

T. Komura • T. Ikeda • K. Hoshino • A. Shibamura
Department of Interdisciplinary Studies for Advanced Aged Society, Graduate School of Human Life Science, Osaka City University, Osaka 558-8585, Japan

Y. Nishikawa (✉)
Department of Interdisciplinary Studies for Advanced Aged Society,
Graduate School of Human Life Science, Osaka City University, 3-3-138 Sugimoto,
Sumiyoshi-ku, Osaka 558-8585, Japan
e-mail: nisikawa@life.osaka-cu.ac.jp

E. Mylonakis et al. (eds.), *Recent Advances on Model Hosts*,
Advances in Experimental Medicine and Biology 710, DOI 10.1007/978-1-4419-5638-5_3,

the worm has the advantage of a short and reproducible life span (Finch and Ruvkun 2001). Recently, after Ausubel et al. reported infection due to *Pseudomonas aeruginosa* (Kurz and Tan 2004; Nicholas and Hodgkin 2004; Tan et al. 1999), the organism has also been recognized as an alternative to mammalian models of infection with bacterial pathogens. In the field of innate immunity research, *C. elegans* is becoming one of the most important experimental animals, similar to the fruit fly *Drosophila* (Kurz and Tan 2004; Nicholas and Hodgkin 2004; Schulenburg et al. 2004).

Age at infection is one of the most important determinants of disease morbidity and mortality (Miller and Gay 1997). Because aging is accompanied by functional and metabolic alterations in cells and tissues, senescence of the immune system results in an age-related increase of infections, malignancy, and autoimmunity (Grubeck-Loebenstein 1997; Moulias et al. 1985). Elderly humans have increased mortality from many different types of infections (Bradley and Kauffman 1990).

Whether nutritional control can retard senescence of immune function and decrease mortality from infectious diseases has not yet been established; the difficulty of establishing a model has made this a challenging topic to investigate. Although some studies have shown successful improvement of biomarkers relating to immunological functions (Bogden and Louria 2004), few reports have shown a beneficial influence of nutrition on immunity and the resultant outcome of experimental infection (Hayek et al. 1997; Effros et al. 1991). Hence we have studied whether *C. elegans* is a useful model host in the fields of immunosenescence and immunonutrition.

C. elegans as a Model for Immunonutrition

Probiotic bacteria are defined as living microorganisms that exert beneficial effects on human health when ingested in sufficient numbers (Naidu et al. 1999). Lactic acid bacteria (LAB) have been used in various fermented foods since antiquity. Metchnikoff, who first proposed the concept of probiotic bacteria in 1907, hypothesized that lactobacilli were important for human health and longevity (Metchnikoff 1907). LAB are the most commonly used probiotic microorganisms. LAB have been found to have a variety of physiological influences on their hosts, including antimicrobial effects, microbial interference, supplementary effects of nutrition, anti-tumor effects, reduction of serum cholesterol and lipids, and immunomodulatory effects. However, there have been no reports concerning the influence of LAB on longevity and immunosenescence.

First, we evaluated whether LAB could contribute to host defenses and prolong the lifetime of *C. elegans* (Ikeda et al. 2007). Lactobacilli and bifidobacteria were fed to worms, and their life span and resistance to *Salmonella enterica* were compared with those of worms fed *Escherichia coli* OP50, an international standard food for *C. elegans*. The worms were generally infected with inocula on conventional nematode growth medium, which contains peptone, raising the possibility that the inoculated pathogen would have proliferated regardless of whether it could successfully infect the nematodes and derive nutrition from the hosts. Garsin et al. showed that nutrition available in agar plates does influence the virulence of pathogens on the media (Garsin et al. 2001). Furthermore, some pathogens produce toxic metabolites on nutrient medium in situ (Anyanful et al. 2005). To avoid such a condition, our experiments were performed on modified nematode growth medium (mNGM) containing no peptone as we reported before (Hoshino et al. 2008). Worms fed heat-killed OP50 reportedly live longer than those fed alive bacteria on nutrient NGM, however this difference was not observed on modified NGM.

Feeding nematodes bifidobacteria or lactobacilli resulted in increases in average life span of the nematodes compared to those fed OP50 (Fig. 3.1). To examine whether or not the beneficial effects of LAB were brought about by their harmless nature compared to OP50, survival was compared with that of nematodes fed on heat-killed OP50. Heat-killed OP50 did not prolong the worms' longevity as much as LAB did (Fig. 3.2).

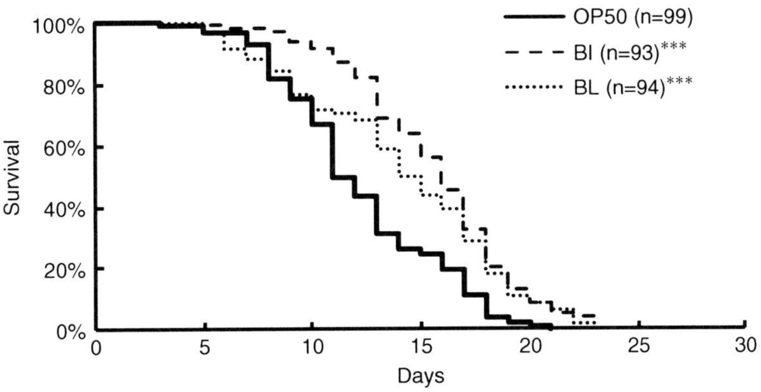

Fig. 3.1 Effects of lactic acid bacteria on the life span of *C. elegans*. Adult worms fed a diet of *E. coli* strain OP50 for 3 days after hatching were transferred to diets of bifidobacteria. The bifidobacteria used were *B. infantis* (BI) or *B. longum* (BL). The life spans of nematodes fed bifidobacteria were significantly extended (***$p<0.001$). The mean life spans (in days) of worms fed *B. infantis* or *B. longum* were 15.1 ± 0.40 (29%) and 13.6 ± 0.50 (17%), respectively: numbers in parentheses are percentage differences in the mean relative to controls fed OP50

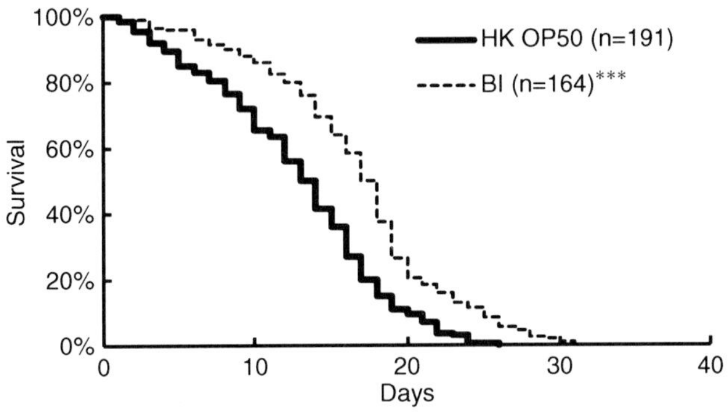

Fig. 3.2 The mean life span (16.3 ± 0.47 days) of worms fed *B. infantis* was prolonged by 33% over that seen with worms fed heat-killed OP50 (12.27 ± 0.42 days) (***$p<0.001$)

Salmonella killed about 40% of the nematodes in 5 days after the worms were transferred to the lawn of this pathogen at age 7 days, while 80% of the worms fed OP50 remained alive after 5 days. The 3-day-old worms were not killed in 5 days when fed either OP50 or *Salmonella*. The 3-day-old worms were clearly more resistant to *Salmonella* compared to the 7-day-old nematodes (Fig. 3.3); the initial number of *Salmonella* recovered from those worms in which infection started at age 3 days was smaller than the number recovered from worms infected from age 7 days (Fig. 3.4).

Importantly, 7-day-old nematodes fed bifidobacteria or lactobacilli from age 3 days were clearly more tolerant to subsequent *Salmonella* infection compared with nematodes fed OP50 before the *Salmonella* infection (Fig. 3.5). LAB seem to make the worms tolerant rather than resistant to *Salmonella* infection; the number of *Salmonella* recovered from worms fed LAB was the same as that recovered from worms grown on OP50.

The mechanisms how LAB brought the worms longevity effects and made them tolerant have not been elucidated. However, if the increased longevity was due to enhancement of host defenses

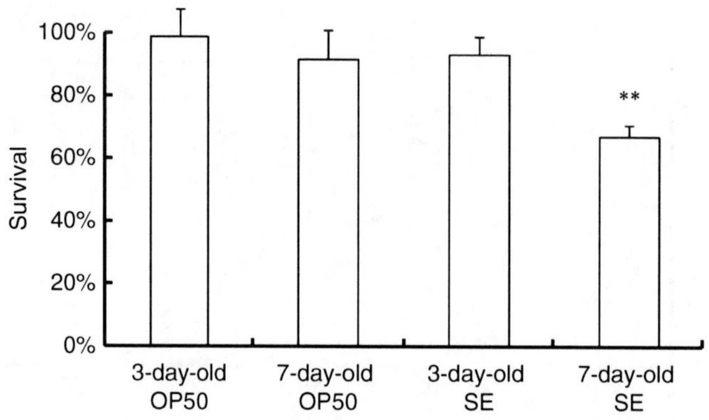

Fig. 3.3 Survival rate at the fifth day of *Salmonella enterica* serovar *Enteritidis* (SE) infection. The death rate of nematodes infected at age 7 days was greater than that of worms infected at age 3 days (**p<0.01). All results are presented as the mean ± standard error of the mean

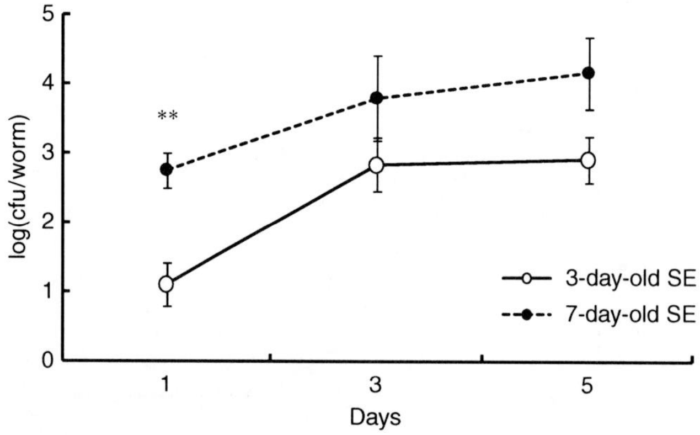

Fig. 3.4 The number of *Salmonella* recovered from young nematodes on the first day after the infection was significantly lower than the number recovered from worms infected at age 7 days (**p<0.01). All results are presented as the mean ± standard error of the mean

as one of the probiotic effects, the worm could be a unique model to study effects of not only LAB but other nutrients on host defense, so-called immunonutrition. *C. elegans* is useful for studying the relationship between innate immunity and pathogens because the nematode lacks an adaptive immune system. Although *C. elegans* does not have phagocytes specialized for innate host defense, it produces a variety of humoral antibiotic substances such as lysozymes, caenopores, lipase, lectins and C3-like thioester-containing proteins, and defensin-like antibiotic peptides. These substances in the bacteriophagous nematodes might be considered to be digestive enzymes; the worm's intestine could be considered analogous to a phagosome. Bacteria resistant to these antibacterial substances are more likely to be nematocidal. Consequently, *C. elegans* may be most suitable to study anti-innate immunity properties of pathogens since the organisms have to contend with the humoral defense factors produced by the host.

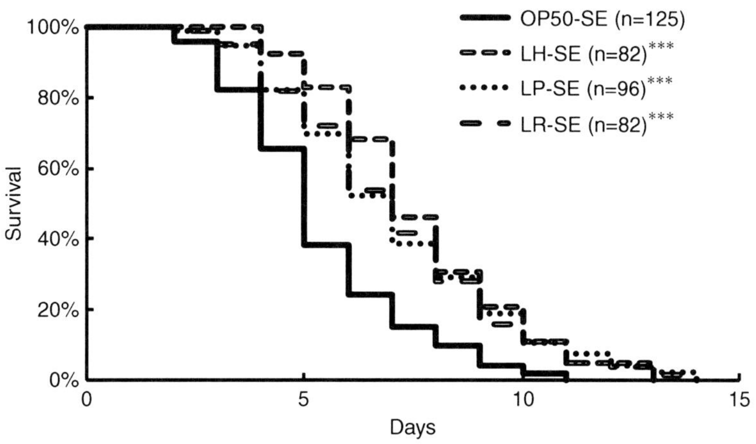

Fig. 3.5 Effects of lactic acid bacteria on resistance of *C. elegans* to *Salmonella*. Adult worms fed a diet of *E. coli* strain OP50 for 3 days after hatching were transferred to a diet of lactobacilli. The lactobacilli used were *L. helveticus* (LH), *L. plantarum* (LP), or *L. rhamnosus* (LR). Four days later the nematodes were transferred to *Salmonella* plates, and survival curves were determined. Nematodes fed each type of lactobacilli were significantly more resistant than controls to the pathogen (***$p < 0.001$). Mean days of survival of worms fed LH, LP, LR before the salmonella infection were 7.1 ± 0.25 (46%), 6.6 ± 0.28 (35%), and 6.6 ± 0.27 (35%), respectively; numbers in parentheses are percentage differences in the mean relative to controls fed OP50

C. elegans as a Model for Immunosenescence

In a second step, we attempted to establish an immunosenescence model using *C. elegans*. Although *Salmonella enterica* serovar Enteritidis killed old nematodes more quickly after the infection than young worms as we described above, we wanted to develop a model capable of testing whether opportunistic infections increase due to immunosenescence. *Legionella pneumophila*, an environmental bacterium naturally found in fresh water, is the major causative agent of Legionnaires' disease (Fields et al. 2002). Fresh water amoeba, a natural host of *Legionella*, has been used as an infection model to study invasion of *Legionella* into human macrophages and subsequent intracellular growth (Jules and Buchrieser 2007). However, analyses using these protozoa have inevitably concentrated on the intracellular lifestyle of *L. pneumophila*. The fate of *Legionella* organisms in non-mammalian metazoans had not been reported (Hilbi et al. 2007) until a very recent report by Brassinga et al. (Brassinga et al. 2010). Since *Legionella* is prone to infect elderly people, we focused on the effects of worm age on *Legionella* infection and the prevention of infection by immunonutrition (Komura et al. 2010). Infections in young and old nematodes were compared. Furthermore, survival curves were compared between worms fed with OP50 and those fed bifidobacteria prior to infection with *Legionella* organisms, since lactic acid bacteria exert beneficial effects on human and animal health (Naidu et al. 1999).

No significant differences in survival were seen between 3-day-old worms fed OP50 and 3-day-old worms infected with virulent *Legionella* strains. However, when the worms were infected from 7.5 days after hatching, the virulent *Legionella* strains were obviously nematocidal (Fig. 3.6). These data show that *L. pneumophila* is virulent even on peptone-free mNGM if the targets are elderly worms. Our previous study showed that *Salmonella* is clearly virulent to both older and younger worms, although more so in elderly worms (Ikeda et al. 2007). These findings appear to be similar to the epidemiological characteristics of both pathogens in humans: *Legionella* tends to infect older people in an opportunistic manner, while *Salmonella* can cause enteritis irrespective of host age.

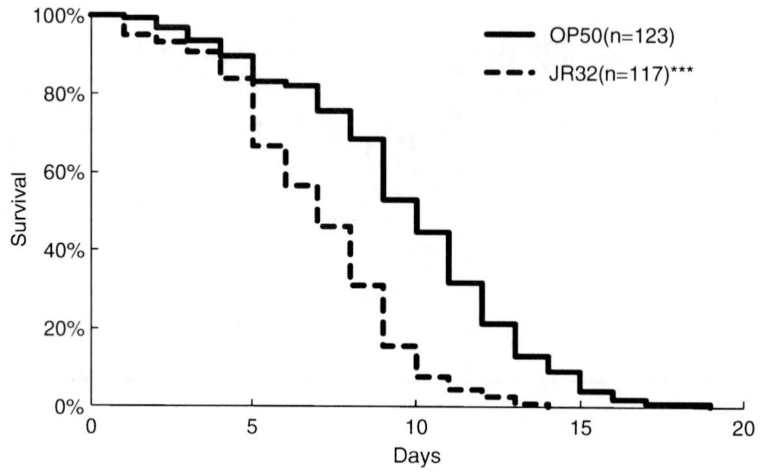

Fig. 3.6 Survival of nematodes infected with *L. pneumophila*. From 7.5 days of age, nematodes were transferred to agar plates covered with the *L. pneumophila* virulent strain JR32. The survival curves were compared with that of worms fed on OP50 (***p<0.001)

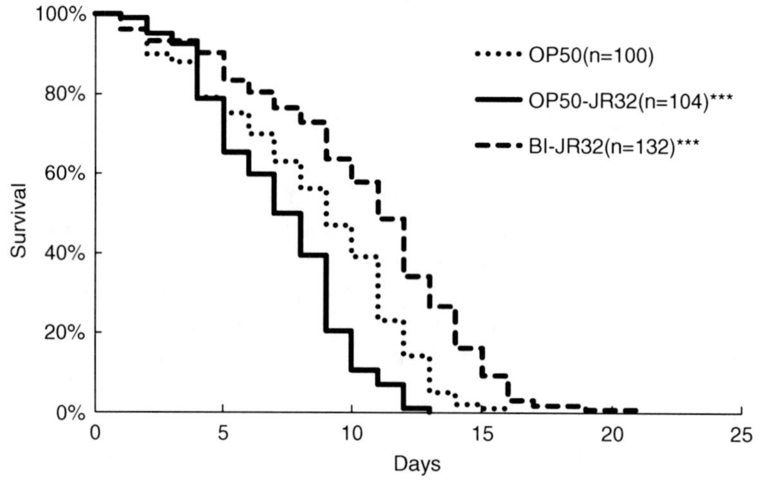

Fig. 3.7 Adult worms fed a diet of bifidobacteria from 3 days of age for 5 days were transferred to *Legionella* plates for infection, and survival curves were drawn. Nematodes fed bifidobacteria were significantly more tolerant than controls to the pathogen (***p<0.001)

As with the case of many other pathogens in the *C. elegans* model, *Legionella* mutants that are less virulent in the lungs of guinea pigs (Miyamoto et al. 2003) or in human macrophages (Sadosky et al. 1993), are also less virulent in *C. elegans*. Interestingly, the attenuated mutant LELA 1718, which is reportedly cytolethal compared to the other attenuated mutants in a cytotoxicity assay with HL-60-derived human macrophages (Sadosky et al. 1993), showed modest virulence in the nematode compared to other avirulent mutants. The pathogenicity of *L. pneumophila* in *C. elegans* seems to correlate well with that in macrophages, and the nematode could serve as a unique host of *Legionella* spp.

Nematodes fed with bifidobacteria prior to *Legionella* infection were resistant to *Legionella* (Fig. 3.7). The number of *Legionella* recovered from the worms showed no significant difference between groups fed with bifidobacteria or OP50. This phenomenon is similar to the tolerance that we observed when nematodes were fed lactic acid bacteria prior to *Salmonella* infection (Ikeda et al. 2007).

Development of a Method for Oral Administration to Nematodes

Despite increased use of *C. elegans* in a variety of studies, there is no efficient method to administer chemicals orally. When chemicals need to be administered to nematodes, they are either dissolved in the NGM or the solution is poured onto the OP50 lawn. *C. elegans* ingests relatively large particles such as bacteria, that are suspended in water, and then spits out much of the liquid, while retaining the particles (Avery and Thomas 1997). This feeding behavior is likely to be inefficient for ingestion of solutions.

In the third step of our studies, we aimed to develop methods for oral administration that is essential for developing a biologically relevant *C. elegans* immunonutrition model. We hypothesized that nematodes would be able to take up liposomes, similar to their ingestion of bacteria. We used liposomes loaded with the hydrophilic fluorescent reagent uranin to test oral administration of water-soluble substances to *C. elegans*, and compared the efficiency of liposome-mediated delivery with conventional methods (Shibamura et al. 2009).

Dietary supplements of antioxidants were previously reported to have positive effects on longevity, while other studies reported controversial results. Water-soluble antioxidants were administered using both our newly developed liposome method and conventional methods to compare the effect on lifespan of nematodes and on host defense against *Salmonella* infection. Using our the liposome method, we showed marked longevity effects of antioxidants on the lifespan of *C. elegans* (Fig. 3.8). Oral administration was more than 200 times as efficient as the conventional method in dose response tests. We expect that this method could open new phase of *C. elegans* research as a model host.

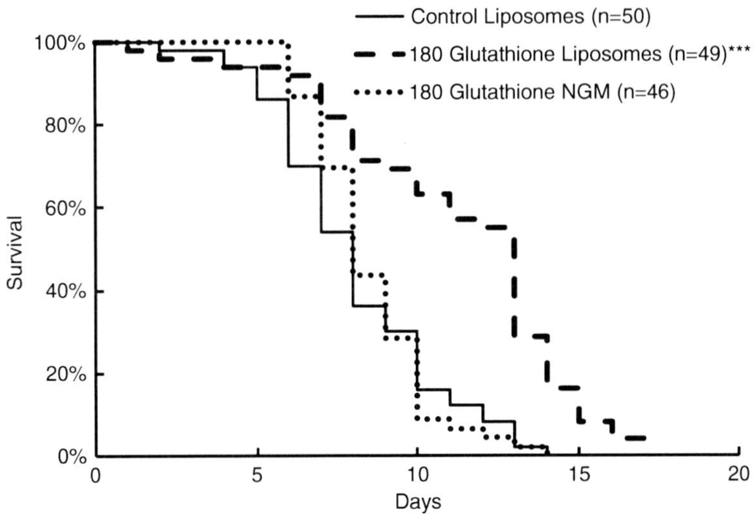

Fig. 3.8 Survival curves of nematodes supplemented with 25 μL of liposomes containing 180 μg reduced glutathione. After hatching, nematodes were grown on *E. coli* OP50 for 3 days, and then the adult worms were divided into groups that were supplemented with chemicals. Water-containing liposomes were administered to control worms and those maintained on mNGM containing 180 μg of glutathione. ***$p<0.001$, compared to the control

Conclusion

Due to increasing ethical considerations as well as economic reasons, the use of mammalian hosts is decreasing in popularity. We showed that *C. elegans* could act as a unique alternative host for immunosenescence and resultant opportunistic infection and immunonutrition experiments. Compared with murine infection models, it is not easy to extrapolate whether the nematocidal activity of a particular pathogen would be reflected in virulence in human pathogenesis. However, for simplicity, transparency, ease of cultivation, and suitability for genetic analysis, *C. elegans* is a uniquely useful model. Particularly for studies on aging of host defense, the worm has the great advantage of a short and reproducible life span.

References

Anyanful A et al (2005) Paralysis and killing of *Caenorhabditis elegans* by enteropathogenic *Escherichia coli* requires the bacterial tryptophanase gene. Mol Microbiol 57:988–1007

Avery L, Thomas JH (1997) Feeding and defecation. In: Riddle DL, Blumenthal T, Meyer BJ, Priess JR (eds) *C. elegans* II. Cold Spring Harbor Laboratory Press, Cold Spring Harbor, pp 679–716

Bogden JD, Louria DB (2004) Nutrition and immunity in the elderly. In: Hughes DA, Gail Darlington J, Bendich A (eds) Diet and human immune function nutrition and health. Humana Press, Totowa, pp 79–101

Bradley SF, Kauffman CA (1990) Aging and the response to salmonella infection. Exp Gerontol 25:75–80

Brassinga AK et al (2010) *Caenorhabditis* is a metazoan host for *Legionella*. Cell Microbiol 12:343–361

Effros RB, Walford RL, Weindruch R, Mitcheltree C (1991) Influences of dietary restriction on immunity to influenza in aged mice. J Gerontol 46:B142–147

Fields BS, Benson RF, Besser RE (2002) *Legionella* and legionnaires' disease: 25 years of investigation. Clin Microbiol Rev 15:506–526

Finch CE, Ruvkun G (2001) The genetics of aging. Annu Rev Genomics Hum Genet 2:435–462

Garsin DA et al (2001) A simple model host for identifying Gram-positive virulence factors. Proc Natl Acad Sci USA 98:10892–10897

Grubeck-Loebenstein B (1997) Changes in the aging immune system. Biologicals 25:205–208

Hayek MG et al (1997) Vitamin E supplementation decreases lung virus titers in mice infected with influenza. J Infect Dis 176:273–276

Hilbi H, Weber SS, Ragaz C, Nyfeler Y, Urwyler S (2007) Environmental predators as models for bacterial pathogenesis. Environ Microbiol 9:563–575

Hoshino K et al (2008) Evaluation of *Caenorhabditis elegans* as the host in an infection model for food-borne pathogens. Jpn J Food Microbiol 25:137–147

Ikeda T, Yasui C, Hoshino K, Arikawa K, Nishikawa Y (2007) Influence of lactic acid bacteria on longevity of *Caenorhabditis elegans* and host defense against *Salmonella entetica* serovar Enteritidis. Appl Environ Microbiol 73:6404–6409

Jules M, Buchrieser C (2007) *Legionella pneumophila* adaptation to intracellular life and the host response: clues from genomics and transcriptomics. FEBS Lett 581:2829–2838

Komura T, Yasui C, Miyamoto H, Nishikawa Y (2010) *Caenorhabditis elegans* as an alternative model host for *Legionella pneumophila* and the protective effects of *Bifidobacterium infantis*. Appl Environ Microbiol 76:4105–4108

Kurz CL, Tan MW (2004) Regulation of aging and innate immunity in *C. elegans*. Aging Cell 3:185–193

Metchnikoff E (1907) The prolongation of life. Heinemann, London

Miller E, Gay N (1997) Effect of age on outcome and epidemiology of infectious diseases. Biologicals 25:137–142

Miyamoto H, Yoshida S, Taniguchi H, Shuman HA (2003) Virulence conversion of *Legionella pneumophila* by conjugal transfer of chromosomal DNA. J Bacteriol 185:6712–6718

Moulias R et al (1985) Respective roles of immune and nutritional factors in the priming of the immune response in the elderly. Mech Ageing Dev 31:123–137

Naidu AS, Bidlack WR, Clemens RA (1999) Probiotic spectra of lactic acid bacteria (LAB). Crit Rev Food Sci Nutr 39:13–126

Nicholas HR, Hodgkin J (2004) Responses to infection and possible recognition strategies in the innate immune system of *Caenorhabditis elegans*. Mol Immunol 41:479–493

Riddle DL, Blumenthal T, Meyer BJ, Priess JR (1997) *C. elegans* II. Cold Spring Harbor Laboratory Press, Cold Spring Harbor

Sadosky AB, Wiater LA, Shuman HA (1993) Identification of *Legionella pneumophila* genes required for growth within and killing of human macrophages. Infect Immun 61:5361–5373

Schulenburg H, Kurz CL, Ewbank JJ (2004) Evolution of the innate immune system: the worm perspective. Immunol Rev 198:36–58

Shibamura A, Ikeda T, Nishikawa Y (2009) A method for oral administration of hydrophilic substances to *Caenorhabditis elegans*: effects of oral supplementation with antioxidants on the nematode lifespan. Mech Ageing Dev 130:652–655

Tan MW, Mahajan-Miklos S, Ausubel FM (1999) Killing of *Caenorhabditis elegans* by *Pseudomonas aeruginosa* used to model mammalian bacterial pathogenesis. Proc Natl Acad Sci USA 96:715–720

Chapter 4
Bacterial Effectors: Learning on the Fly

Laurent Boyer, Nicholas Paquette, Neal Silverman, and Lynda M. Stuart

Abstract A common defining characteristic of pathogenic bacteria is the expression of a repertoire of effector molecules that have been named virulence factors. These bacterial factors include a variety of proteins, such as toxins that are internalized by receptors and translocate across endosomal membranes to reach the cytosol, as well as others that are introduced directly into the cell by means of bacterial secretory apparatuses. Given the importance of these effectors for understanding bacterial pathogenicity, significant effort has been made to dissect their molecular mechanisms of action and their respective roles during infection. Herein we will discuss how *Drosophila* have been used as a model system to study these important microbial effectors, and to understand their contribution to pathogenicity.

Introduction

Over the past two decades a number of findings made in *Drosophila melanogaster* have provided important new insights into mammalian innate immunity (Hoffmann et al. 1999; Martinelli and Reichhart 2005). The power of this system is best exemplified by the discovery that Toll, a receptor used for dorso-ventral patterning in the developing embryo, is reused in the adult fly as a component of a microbial sensing pathway (Lemaitre et al. 1996). This seminal discovery led to identification of the Toll-like receptors (TLRs) as the critical innate immune receptors in mammals (Janeway and Medzhitov 2002). Other such examples demonstrate an amazing conservation between how flies and mammals fight infectious agents (Hoffmann et al. 1999; Akira et al. 2006). The majority of the studies to date have used *Drosophila* as a model host to understand the signals downstream of the two key pattern recognition receptors, Toll and Peptidoglycan Recognition Protein (PGRP)-LC (Lemaitre and Hoffmann

L. Boyer (✉)
INSERM, U895, Centre Méditerranéen de Médecine Moléculaire, Toxines Microbiennes dans la relation hôte pathogènes, Cedex 3, F-06204 Nice, France

Université de Nice-Sophia-Antipolis, UFR Médecine, Nice, France
e-mail: laurent.boyer@unice.fr

N. Paquette • L.M. Stuart
Developmental Immunology, Massachusetts General Hospital/Harvard Medical School, 55 Fruit Street, Boston, MA 02114, USA

N. Silverman
Division of Infectious Disease, Department of Medicine, University of Massachusetts Medical School, 364 Plantation Street, Worcester, MA 01605, USA

E. Mylonakis et al. (eds.), *Recent Advances on Model Hosts*,
Advances in Experimental Medicine and Biology 710, DOI 10.1007/978-1-4419-5638-5_4,

2007; Cherry and Silverman 2006). Similarly, work in mammals has also focused on dissecting Pattern Recognition Receptor (PRR)-triggered pathways (Kawai and Akira 2009). These receptors recognize critical microbial components such as peptidoglycan or LPS, which are found on commensal and pathogenic microbes (Akira et al. 2006). However, a poorly understood aspect of innate immunity is how we differentiate pathogens and non-pathogens. A common defining characteristic of pathogenic bacteria is the expression of effector molecules or so-called 'virulence factors,' which modify host defense mechanisms (Hacker and Kaper 2000; Brodsky and Medzhitov 2009; Finlay and McFadden 2006). These bacterial factors include a variety of proteins, such as toxins that are internalized by receptors and translocate across endosomal membranes to reach the cytosol as well as others that are introduced directly into the cell by means of bacterial secretory apparatuses (Boquet and Lemichez 2003; Henkel et al. 2010). In this chapter we will discuss how *Drosophila* have been used as a system to study these important microbial effectors, and to understand how they contribute to pathogenicity.

Microbial Effectors

Although the term 'effector' is sometimes used only to describe the molecules introduced by the type III secretory apparatus expressed primarily by Gram-negative microbes, for simplicity, we will use this term more loosely to encompass all secreted toxins. Effectors manipulate a variety of processes, including innate immune signaling pathways, the cytoskeleton, protein translation, ubiquitination and the cell cycle (Boquet and Lemichez 2003; Ribet and Cossart 2010). Although these molecules make important contributions to the pathogenic potential of a microorganism, systematic study is hindered by a number of issues. Firstly, despite targeting a relatively limited number of host cellular functions and processes, they demonstrate a remarkable structural diversity. For this reason it is often difficult to predict their mechanism of action or their cellular targets. Secondly, any particular bacteria can introduce a number of effectors into the host. Importantly, as these effectors are frequently redundant for particular activities, classic mutant/deletion based strategies do not always result in clear phenotypes. Thirdly, they are often toxic to eukaryotic cells especially when ectopically expressed, limiting the work that can be done *in vitro*. Thus, the study of each specific effector has its unique challenges. Aside from the many powerful genetic tools available in the fly system and the well-characterized innate immune pathways, *Drosophila* offers a number of advantages to study these types of molecules. For example, in tissue culture, the tightly regulated *Drosophila* metallothionein promoter is ideal for expression of potentially toxic effector proteins, which may kill cells from the leaky expression found on other promoters. Similarly, it is possible to use the UAS system driven by Gal4, with or without the addition of Gal80 suppressor, to achieve tight in vivo tissue-specific or inducible expression. The wide range of tools available lead us to suggest that *Drosophila* may be an attractive system in which to try and better understand effectors and their mechanisms of action. Here we will not attempt to provide a comprehensive review but rather discuss a few examples in which *Drosophila* has already been used to study bacterial effectors and provide the proof of principle for this approach. We will then discuss some potential future directions and applications of this as a model system.

Using Drosophila to Study Bacterial Effectors that Regulate Rho GTPases: Filling in the GAPs

More than 30 bacterial effectors from Gram-negative or Gram-positive bacteria directly or indirectly target the most studied Rho GTPase members : Rho, Rac and/or Cdc42 (Boquet and Lemichez 2003; Visvikis et al. 2010). RhoGTPases are pleiotropic regulators of cellular homeostasis, and are more specifically involved in the regulation of the cytoskeletal rearrangements necessary for migration or phagocytosis (Etienne-Manneville and Hall 2002). Therefore, Rho GTPases are not only master regulators of the cytoskeleton but also central elements of the host responses against pathogens (Bokoch 2005). For this reason, modification of the host Rho GTPases is a widespread strategy used by bacterial pathogens to manipulate mammalian host defenses, and they are frequently targeted by bacterial virulence factors (Boquet and Lemichez 2003). RhoGTPases cycle between an active GTP-bound state and an inactive GDP bound state. Their activation requires guanine nucleotide exchange factors (GEF), whereas GTPase-activating proteins (GAP) stimulate GTP hydrolysis to inactivate the RhoGTPases (Etienne-Manneville and Hall 2002). Bacteria have evolved strategies to target the RhoGTPases family either by direct post-translational modification or by mimicking GEF or GAP activity (Boquet and Lemichez 2003; Stebbins and Galan 2001). Many of the bacterial effectors isolated from pathogenic bacteria are inhibitors of RhoGTPases. These bacterial proteins are either used to disrupt the RhoGTPase cycle or to block the binding of these molecules to their downstream effectors. As highlighted by Boquet and Lemichez (2003), it is surprising to observe that Rac GTPase seems to be the only common target of this group of bacterial toxins, and the fact that Rac regulates numerous cellular pathogen defense pathways is probably not a coincidence. Among these bacterial RhoGTPase inhibitors is a family of bacterial effectors that triggers GTP hydrolysis to inhibit RhoGTPases, thus mimicking eukaryotic GAP proteins. These bacterial effectors, including SptP from *Salmonella typhimurium*, YopE from *Yersinia spp.* and ExoS from *Pseudomonas aeruginosa,* are prokaryotic GAP proteins. These three bacterial proteins have a GAP domain that shares no sequence similarities but nonetheless all have potent GAP activity (Stebbins and Galan 2001). This suggests that these bacterial effectors are the product of a convergent evolution and that many microbes have evolved distinct strategies to inhibit a common target, the RhoGTPases.

Marie Odile Fauvarque's work on ExoS toxin from *Pseudomonas aeruginosa* provides one of the first, and most elegant examples, of how *Drosophila* can be used to investigate the effect of a bacterial toxin (Avet-Rochex et al. 2005). ExoS is a *P. aeruginosa* exotoxin directly translocated into the host cell cytoplasm through the type III secretion system. ExoS contains a GAP domain that prevents cytoskeleton reorganization by the Rho family of GTPases and an ADP-ribosyltransferase domain that modifies RasGTPases (Aktories et al. 2000). To investigate the role of the GAP domain of ExoS toxin, they took advantage of the genetically tractable fly system to generate a transgenic *Drosophila* expressing the ExoS GAP domain (ExoSGAP) of the toxin. Through transgenic expression, they were able to identify Rac (rather than Rho or Cdc42) as the in vivo target of this effector. Moreover, using this system they showed that flies resistance to *P. aeruginosa* infections was altered when ExoSGAP was expressed either ubiquitously or specifically in hemocytes, but not when expressed in the fat body, the major source of anti-microbial peptide production (Avet-Rochex et al. 2005). This suggested that the innate immune response is not dependent on a modified anti-microbial peptide production. Flies expressing ExoSGAP showed increased sensitivity to infection with Gram-positive *Staphylococcus aureus,* which was attributed to the reduced phagocytic capacity of ExoSGAP-expressing hemocytes (Avet-Rochex et al. 2005). This system allowed the authors to decipher in vivo the role of the GAP domain of ExoS on phagocytosis, and to suggest a major role of ExoS to inhibit cellular defence during infection with *P. aeruginosa*.

Interestingly, this virulence strategy is not specific to bacteria and has also been utilized by eukaryotic parasites to corrupt the host response (Colinet et al. 2007). The parasitoid wasp *Leptopilina boulardi* is a natural parasite to *Drosophila* larvae and recently, LbGAP, a GAP parasite-derived protein specific for the Rac GTPases, has been shown to translocate from the parasite into *Drosophila* hemocytes. Similar to the role of bacterial RhoGAP in mammals, this could be a protection mechanism used by the parasitoid wasp to be protected from the innate immune response of *Drosophila* host larvae (Colinet et al. 2007). From a pathogen point of view, these observations highlight the fact that pathogens that infect insects and mammals use evolutionarily convergent strategies, targeting the same key factors to control the host response of their particular hosts. From the host perspective, these observations indicate a conserved and important role of the Rac GTPase in the innate immunity of flies to humans.

Effector-Triggered Immunity: PR1 and CNF1, Alerting the Host to the Presence of Pathogens

Current models of innate immunity suggest that responses are triggered primarily by pattern recognition receptors (PRRs) that recognize conserved molecular patterns expressed by microbes (Janeway and Medzhitov 2002). Although most PRR ligands are shared between commensal and virulent strains, the host demonstrates a remarkable capacity to tailor the response to the virulence of the invading microorganism. However, how the host can detect specifically the virulence associated with microbes is poorly understood. One possibility, suggested from work in plants, is that effectors themselves can be sensed by the immune system. In resistant plants, such effectors are able to induce protective immune responses, that are referred to in the plant field as 'effector-triggered immunity' (Jones and Dangl 2006). Although suggested from plants, there are very few studies in mammals that have addressed this. However, two studies in flies suggest that effector-triggered immunity is an important mechanism for discerning pathogenic microbes by metazoans.

Drosophila immunity to fungal and Gram-positive pathogens is dependent largely on the Toll signaling pathway (Lemaitre and Hoffmann 2007). The canonical activation step in this pathway is the cleavage of the secreted protein Spatzle. Once cleaved, Spatzle then acts as the Toll ligand, inducing multimerization and signaling similar to mammalian MyD88-dependent NF-kB activation (Weber et al. 2007). To understand how this pathway was activated by virulent microbes, Gottar et al. (2006) studied the response to an entomopathogenic fungi *Metarhizium anisopliae* in flies. One of the main virulence strategies used by this fungus is mediated by PR1, a member of the subtilisin family of proteases that perforate the cuticle barrier and allow entry of the fungi into the insect body cavity (Clarkson and Charnley 1996). To investigate the contribution of the PR1 protease the authors generated PR1 transgenic flies. Surprisingly, ectopic expression of this protease was sufficient to drive an immune response, and these flies had increased expression of Drosomycin in the absence of immune challenge. The mechanism involved the ability of PR1 to initiate a cascade of events resulting in Persephone-dependent Drosomycin expression. From this elegant experiment they propose a model where sensing of this fungus is mediated by a dual detection system; the first triggered by recognition of the fungal cell wall, the second in response to the secreted virulence factor, and both are required to maximally activate the Toll pathway (Gottar et al. 2006).

An interesting extension of the work of Gottar et al. (2006), is that the immune response induced by bacterial effectors might actually contribute to protective immunity, and as in plants, might help the resistant host limit bacterial replication. In our recent work we have used flies to address this possibility (Boyer et al. 2011). We have focused on Cytotoxic Necrotizing Factor 1 (CNF1), a toxin from uropathogenic *Escherichia coli.* CNF1 is an archetypal example of a RhoGTPase activating toxin and belongs to a family including CNF2 from *E. coli* as well as DNT from *Bordetella* spp. or CNFy from *Yersinia pseudotuberculosis* (Lemonnier et al. 2007). CNF1 is a deamidase, which catalyzes the activation of RhoGTPases (Flatau et al. 1997; Schmidt et al. 1997). CNF1 intoxication of mammalian epithelial cells induced activation of Rac. This in turn is involved in the clustering of different components of the SCF ubiquitylation complex, comprising Skp1 and neddylated-Cullin-1, together with IkBa and is associated with NF-kB p65 translocation to the nucleus (Boyer et al. 2004). More recently, we have used *Drosophila* to identify the innate immune pathway initiated in response to the CNF1 toxin (Boyer et al. 2011). We found that CNF1 toxin is sufficient to initiate defense signals in the absence of other bacterial components, and identified a conserved immune pathway that signals initiation of this response in flies and mammals. Analogous to 'effector-triggered immunity' observed in plants (Jones and Dangl 2006), we propose that the inappropriate activation of RhoGTPases by CNF1 is effectively monitored by the host, to the detriment of the bacteria. This mechanism of immune surveillance, based on monitoring the activity of virulence factors, provides a framework for a recognition system able to deal with the large number of highly varied microbial toxins targeting RhoGTPases. We anticipate that other targets of microbial virulence determinants will be similarly monitored. This work provides the first example of an evolutionarily conserved means by which pathogenicity is detected through sensing a microbial effector.

Using Drosophila to Study Effectors that Inhibit Innate Immune Responses: Yersinia pestis YopJ

Known primarily as a pathogen of historical importance and the causative agent of 'plague,' *Yersinia pestis* is a highly virulent bacterium. To reach its pathogenic potential, during an infection *Y. pestis* injects a number of bacterial effector proteins directly into host immune cells using a type III secretion system (Cornelis and Yersinia 2002).These effector proteins function to inhibit various cellular and immune pathways. Recently the precise function of one of these proteins, YopJ, has been debated. YopJ was first observed to promote apoptosis and inhibit NF-kB signaling pathways, which are essential for innate immune activation (Monack et al. 1997; Palmer et al. 1998, 1999). Initially YopJ was proposed to act as an ubiquitin-like protein protease, cleaving ubiquitin or ubiquitin-like proteins from their conjugated substrates (Orth et al. 2000; Sweet et al. 2007; Zhou et al. 2005). However, recent evidence indicates that YopJ has a novel function, that of a serine/threonine acetyl-transferase (Mittal et al. 2006; Mukherjee et al. 2006). In this role YopJ is proposed to acetylate critical serine and threonine residues of MAP2 kinases such as MKK2, MKK6 and IKK. In order to further understand the molecular role of YopJ, we (Paquette et al. 2011) have used *Drosophila*. Similar work using the YopJ related protein AvrA from *Salmonella typhimurium* has also been described (Jones et al. 2008). Over expression of YopJ in immune stimulated *Drosophila* S2 cells was found to inhibit the IMD pathway, without affecting the Toll pathway, indicating that YopJ has a specific molecular target. Using, RNAi to probe this phenotype, we identified a new target for YopJ, TAK1, a member of the MAP3 kinase family. Thus mechanisms of effector-mediated immune suppression can also be identified with this approach.

Drosophila as a Tool to Decipher the Role of Effectors in Chronic Infection and Inflammation: Helicobacter pylori CagA

In addition to the obvious consequence that bacterial effectors have on regulating innate immunity, there are pleotropic consequences during chronic infection. As an example, chronic *H. pylori* infection is the causative agent of gastritis, peptic ulcers and gastric cancer (Rothenbacher 2003). During infection the bacteria uses a type four secretion system to inject bacterial toxins directly into the host cells. One major virulence factor that associates with *H. pylori* is the cytotoxin-associated gene A (CagA) protein (Bourzac and Guillemin 2005). Once inside a host cell CagA is phosphorylated by Src kinases and acts to disrupt receptor typrosine kinase (RTK) signaling pathways by activating Src homology 2 domain containing tyrosine phosphatase (SHP-2). In tissue culture systems CagA has been shown to interact and activate SHP-2, resulting in cell elongation (Hatakeyama 2006). As SHP-2 normally binds to Gab proteins, CagA is hypothesized to mimic Gab proteins even though they share no sequence similarity, and thus to function as an oncogene by activating RTK signaling. In order to more fully understand the mechanism of CagA in epithelial tissues Botham et al. (2008) undertook a study in which they expressed CagA in the eye of developing *Drosopihla*. CagA expression driven by the GMR driver resulted in a severe eye deformation. In order to determine if CagA could mimic Gab, Botham and collegues performed an elegant rescue experiment using the *Drosophila* Gab homolog, DOS. In homozygous *dos* loss of function mutants, pupal development is severely reduced and adult animals are never generated. Using a ubiquitous drive (Hsp-Gal4), expression of CagA rescued the *dos* mutant lethality. Furthermore, using the FLP/FRT to generate *dos/dos* in the eye, it was also shown that CagA could directly rescue *dos/dos* dependent photoreceptor development. These data show that CagA does in fact act to mimic DOS during eye development. Lastly under the assumption that CagA mimics Gab, it was tested if the SHP-2/CSW protein was required downstream for proper eye development. Using *csw* mutant *Drosophila*, it was shown that overexpression of CagA did not rescue the *csw* dependent lack of photoreceptors, indicating that CagA requires SHP-2/CSW for proper function. Taken together this work shows how the *H. pylori* bacterial effector protein CagA functions as a mimic of Gab in an in vivo epithelial model system, and is an elegant example of what is possible using the powerful genetic tools available in *Drosophila* to investigate the function of bacterial effectors.

Future Directions

Finding the Bad Guys: Using Drosophila to Identify Bacterial Effectors

In work that has been pioneered by Dr Svenja Stöven (Vonkavaara et al. 2008; Ahlund et al. 2010), it has recently been shown that *Drosophila* might be a powerful system to screen for bacterial effectors involved in virulence. To demonstrate that *D. melanogaster* is a suitable in vivo model for the identification of *F. tularensis* virulence determinants, they first targeted the *igl* operon and the regulator *mglA*, bacterial genes known to be required for bacterial intracellular growth and virulence in mice. They injected flies with either the wild type strain or with isogenic Δ*iglB,* Δ*iglC,* Δ*iglD* and Δ*mglA* mutants and found that flies injected with Δ*iglB,* Δ*iglC,* Δ*iglD or* Δ*mglA* mutants survived significantly longer than wild type-infected flies (Vonkavaara et al. 2008). They extended this approach by screening for *F. novicida* genes involved in virulence. They performed a directed screen using an *F. novicida* transposon insertion library, and scored the survival of infected fruit flies (Ahlund et al. 2010). This approach allowed them to identify clusters of genes required for *Francisella* virulence, and established *Drosophila* as a useful in vivo system to identify bacterial genes involved in "virulence" or "avirulence" of pathogenic bacteria.

Look and Learn: Intravital Imaging of Drosophila to Monitor the Consequences of Bacterial Effectors During In Vivo Infection

One of the limits in studying host pathogen interactions is the visualization of the pathogens as they interact with the host cells in vivo, especially during the very early stages of infection. Despite the development of intravital microscopy and luminescence lifetime imaging technology in mice, the resolution and potential for investigation in mammals are limited. Like zebrafish and nematodes, *Drosophila* have proved to be a powerful model for in vivo microscopy, and this has been extensively used to study *Drosophila* early embryonic development and wound repair (Stramer and Wood 2009; Stramer et al. 2005). Recently, intravital imaging has been used to follow bacterial infection in real time. Will Wood's group (Vlisidou et al. 2009) has adapted and developed a powerful imaging system using the *Drosophila* embryo to study the role of a bacterial effector called Makes Caterpillars Floppy (Mcf1), produced by the insect pathogen *Photorhabdus asymbiotica*. Using this model they show that embryonic hemocytes can sense and phagocytose non-pathogenic *Escherichia coli*. However, when embryos were infected with *P. asymbiotica*, hemocytes bind to the bacteria but become immotile 20 min after infection. Using *Drosophila*, Mcf1 toxin was identified as the bacterial effector responsible for this striking phenotype, as embryos injected with *E. coli* producing Mcf1 or purified toxin alone, recapitulate the hemocyte immobilization phenotype (Vlisidou et al. 2009). This study also used *Drosophila* mutants to show that the immobilization phenotype requires the internalization of the Mcf1 toxin, and that this phenotype is dependent of the GTPase Rac. This work was facilitated by the use of the combination of a genetically tractable host, *Drosophila melanogaster*, and a genetically tractable microbe, *E. coli*, to elucidate the role of the Mcf1 toxin during the early steps of infection in vivo (Vlisidou et al. 2009). Moreover, these studies demonstrate that it is possible to obtain subcellular resolution in living organisms, and thus highlight the value of *Drosophila* for live cell imaging and intravital microscopy for the study of the immune response in vivo.

Acknowledgments We thank Emmanuel Lemichez for critical reading of the manuscript. LMS is supported by startup funds from MGHfC, MGH ECOR and grants from NIH/NIAID. NS is supported by grants from the NIH/NIAID (AI060025 and AI074958) and from BWF. NP is supported by a grant from the NIH (U54 AI057159). LB is supported by a fellowship from the Ligue Nationale Contre le Cancer.

References

Ahlund MK, Ryden P, Sjostedt A, Stoven S (2010) Directed screen of Francisella novicida virulence determinants using Drosophila melanogaster. Infect Immun 78:3118–3128

Akira S, Uematsu S, Takeuchi O (2006) Pathogen recognition and innate immunity. Cell 124:783–801

Aktories K, Schmidt G, Just I (2000) Rho GTPases as targets of bacterial protein toxins. Biol Chem 381:421–426

Avet-Rochex A, Bergeret E, Attree I, Meister M, Fauvarque MO (2005) Suppression of Drosophila cellular immunity by directed expression of the ExoS toxin GAP domain of Pseudomonas aeruginosa. Cell Microbiol 7:799–810

Bokoch GM (2005) Regulation, of innate immunity by Rho GTPases. Trends Cell Biol 15:163–171

Boquet P, Lemichez E (2003) Bacterial virulence factors targeting Rho GTPases: parasitism or symbiosis? Trends Cell Biol 13:238–246

Botham CM, Wandler AM, Guillemin K (2008) A transgenic Drosophila model demonstrates that the Helicobacter pylori CagA protein functions as a eukaryotic Gab adaptor. PLoS Pathog 4:e1000064

Bourzac KM, Guillemin K (2005) Helicobacter pylori-host cell interactions mediated by type IV secretion. Cell Microbiol 7:911–919

Boyer L et al (2004) Rac GTPase instructs nuclear factor-kappaB activation by conveying the SCF complex and IkBalpha to the ruffling membranes. Mol Biol Cell 15:1124–1133

Boyer et al (2011) Identification of a conserved mechanism of effector-triggered immunity mediated by IMD and Rip proteins. Immunity in press

Brodsky IE, Medzhitov R (2009) Targeting of immune signalling networks by bacterial pathogens. Nat Cell Biol 11:521–526

Cherry S, Silverman N (2006) Host-pathogen interactions in drosophila: new tricks from an old friend. Nat Immunol 7:911–917

Clarkson JM, Charnley AK (1996) New insights into the mechanisms of fungal pathogenesis in insects. Trends Microbiol 4:197–203

Colinet D, Schmitz A, Depoix D, Crochard D, Poirie M (2007) Convergent use of RhoGAP toxins by eukaryotic parasites and bacterial pathogens. PLoS Pathog 3:e203

Cornelis GR (2002) Yersinia, type III secretion: send in the effectors. J Cell Biol 158:401–408

Etienne-Manneville S, Hall A (2002) Rho GTPases in cell biology. Nature 420:629–635

Finlay BB, McFadden G (2006) Anti-immunology: evasion of the host immune system by bacterial and viral pathogens. Cell 124:767–782

Flatau G et al (1997) Toxin-induced activation of the G protein p21 Rho by deamidation of glutamine. Nature 387:729–733
Gottar M et al (2006) Dual detection of fungal infections in Drosophila via recognition of glucans and sensing of virulence factors. Cell 127:1425–1437
Hacker J, Kaper JB (2000) Pathogenicity islands and the evolution of microbes. Annu Rev Microbiol 54:641–679
Hatakeyama M (2006) The role of Helicobacter pylori CagA in gastric carcinogenesis. Int J Hematol 84:301–308
Henkel JS, Baldwin MR, Barbieri JT (2010) Toxins from bacteria. EXS 100:1–29
Hoffmann JA, Kafatos FC, Janeway CA, Ezekowitz RA (1999) Phylogenetic perspectives in innate immunity. Science 284:1313–1318
Janeway CA, Medzhitov R (2002) Innate immune recognition. Annu Rev Immunol 20:197–216
Jones JD, Dangl JL (2006) The plant immune system. Nature 444:323–329
Jones RM et al (2008) Salmonella AvrA coordinates suppression of host immune and apoptotic defenses via JNK pathway blockade. Cell Host Microbe 3:233–244
Kawai T, Akira S (2009) The roles of TLRs, RLRs and NLRs in pathogen recognition. Int Immunol 21(4):317–337
Lemaitre B, Hoffmann J (2007) The host defense of Drosophila melanogaster. Annu Rev Immunol 25:697–743
Lemaitre B, Nicolas E, Michaut L, Reichhart JM, Hoffmann JA (1996) The dorsoventral regulatory gene cassette spatzle/toll/cactus controls the potent antifungal response in Drosophila adults. Cell 86:973–983
Lemonnier M, Landraud L, Lemichez E (2007) Rho GTPase-activating bacterial toxins: from bacterial virulence regulation to eukaryotic cell biology. FEMS Microbiol Rev 31:515–534
Martinelli C, Reichhart JM (2005) Evolution and integration of innate immune systems from fruit flies to man: lessons and questions. J Endotoxin Res 11:243–248
Mittal R, Peak-Chew SY, McMahon HT (2006) Acetylation of MEK2 and I kappa B kinase (IKK) activation loop residues by YopJ inhibits signaling. Proc Natl Acad Sci USA 103:18574–18579
Monack DM, Mecsas J, Ghori N, Falkow S (1997) Yersinia signals macrophages to undergo apoptosis and YopJ is necessary for this cell death. Proc Natl Acad Sci USA 94:10385–10390
Mukherjee S et al (2006) Yersinia YopJ acetylates and inhibits kinase activation by blocking phosphorylation. Science 312:1211–1214
Orth K et al (2000) Disruption of signaling by Yersinia effector YopJ, a ubiquitin-like protein protease. Science 290:1594–1597
Palmer LE, Hobbie S, Galan JE, Bliska JB (1998) YopJ of Yersinia pseudotuberculosis is required for the inhibition of macrophage TNF-alpha production and downregulation of the MAP kinases p38 and JNK. Mol Microbiol 27:953–965
Palmer LE, Pancetti AR, Greenberg S, Bliska JB (1999) YopJ of Yersnia spp. is sufficient to cause downregulation of multiple mitogen-activated protein kinases in eukaryotic cells. Infect Immun 67:708–716
Ribet D, Cossart P (2010) Post-translational modifications in host cells during bacterial infection. FEBS Lett 584:2748–2758
Rothenbacher D, Brenner H (2003) Burden of Helicobacter pylori and H. pylori-related diseases in developed countries: recent developments and future implications. Microbes Infect 5:693–703
Schmidt G et al (1997) Gln 63 of Rho is deamidated by Escherichia coli cytotoxic necrotizing factor-1. Nature 387:725–729
Stebbins CE, Galan JE (2001) Structural mimicry in bacterial virulence. Nature 412:701–705
Stramer B, Wood W (2009) Inflammation and wound healing in Drosophila. Methods Mol Biol 571:137–149
Stramer B et al (2005) Live imaging of wound inflammation in Drosophila embryos reveals key roles for small GTPases during in vivo cell migration. J Cell Biol 168:567–573
Sweet CR, Conlon J, Golenbock DT, Goguen J, Silverman N (2007) YopJ targets TRAF proteins to inhibit TLR-mediated NF-kappaB, MAPK and IRF3 signal transduction. Cell Microbiol 9:2700–2715
Visvikis O, Maddugoda MP, Lemichez E (2010) Direct modifications of Rho proteins: deconstructing GTPase regulation. Biol Cell 102:377–389
Vlisidou I et al (2009) Drosophila embryos as model systems for monitoring bacterial infection in real time. PLoS Pathog 5:e1000518
Vonkavaara M, Telepnev MV, Ryden P, Sjostedt A, Stoven S (2008) Drosophila melanogaster as a model for elucidating the pathogenicity of Francisella tularensis. Cell Microbiol 10:1327–1338
Weber AN et al (2007) Role of the Spatzle Pro-domain in the generation of an active toll receptor ligand. J Biol Chem 282:13522–13531
Zhou H et al (2005) Yersinia virulence factor YopJ acts as a deubiquitinase to inhibit NF-kappa B activation. J Exp Med 202:1327–1332

Chapter 5
A *Drosophila* Asthma Model – What the Fly Tells Us About Inflammatory Diseases of the Lung

Thomas Roeder, Kerstin Isermann, Kim Kallsen, Karin Uliczka, and Christina Wagner

Abstract Asthma and COPD are the most relevant inflammatory diseases of the airways. In western countries they show a steeply increasing prevalence, making them to a severe burden for health systems around the world. Although these diseases are typically complex ones, they have an important genetic component. Genome-wide association studies have provided us with a relatively small but comprehensive list of asthma susceptibility genes that will be extended and presumably completed in the near future. To identify the role of these genes in the physiology and pathophysiology of the lung, genetically tractable model organisms are indispensable and murine models were the only ones that have been extensively used. An urgent demand for complementary models is present that provide specific advantages lacking in murine models, especially regarding speed and flexibility. Among the model organisms available, only the fruit fly *Drosophila melanogaster* shares a comparable organ composition and at least a lung equivalent. It has to be acknowledged that the fruit fly *Drosophila* has almost completely been ignored as a model organism for lung diseases, simply because it is devoid of lungs. Nevertheless, its airway system shows striking similarities with the one of mammals regarding its physiology and reaction towards pathogens, which holds the potential to function as a versatile model in asthma-related diseases.

Asthma and Other Chronic Inflammatory Diseases of the Lung

Asthma is the most common chronic inflammatory disease of the lung. Hallmarks of the disease include airway obstruction, bronchospasm, wheezing, coughing and shortness of breath. Chronic asthma comprises complex remodelling processes of the airway contributing to the disease phenotype. Like most inflammatory diseases, asthma is a complex disease, meaning that different intrinsic and extrinsic factors contribute to pathogenesis. Intrinsic factors, meaning genetic susceptibility, are

T. Roeder (✉) • K. Isermann
Christian-Albrechts University Kiel, Kiel, Germany
e-mail: troeder@zoologie.uni-kiel.de; k.isermann@web.de

K. Kallsen • K. Uliczka
Research Center Borstel, Borstel, Germany
e-mail: kkallsen@fz-borstel.de; kuliczka@fz-borstel.de

C. Wagner
University Hospital Eppendorf, Hamburg, Germany
e-mail: wagner-christinawagner@web.de

E. Mylonakis et al. (eds.), *Recent Advances on Model Hosts*,
Advances in Experimental Medicine and Biology 710, DOI 10.1007/978-1-4419-5638-5_5,

highly relevant for asthma development, because family and twin studies revealed that the risk of developing asthma is increased several fold in first-degree relatives of asthma patients (Holloway et al. 2008). A number of different asthma susceptibility genes have been identified by positional cloning as well as by candidate gene studies. The results of these studies have challenged the generally accepted view of asthma pathogenesis, leading to the suggestion that new and unexpected pathways are involved in asthma development, which would substantially increase the number of potential targets for pharmacological intervention (Holgate and Polosa 2008). The large number of genome-wide association studies performed with large and well-defined cohorts of patients that are currently being carried out will provide us with an almost complete set of asthma susceptibility genes. The usefulness of this rich source of information is currently limited by our inability to understand the connection between genes and disease development, at least for the majority of them. In asthma, the situation is obviously similar to all other complex diseases: the identification of susceptibility genes is the beginning and not the end of the story. To profit from these rich sources, model organisms are of paramount importance to elucidate the relevance of these genes and the genetic variations identified. Transgenic mice, including "humanized" mice, are currently the sole genetically tractable models used successfully in asthma research (Finkelman and Wills-Karp 2008). However, some drawbacks of these models hinder progress in our understanding of the disease. Although they are very useful, transgenic and mutant mice have drawbacks that obscure a direct analysis of the expected phenotypes. Genetic redundancy and developmental plasticity are especially important issues that often impair the direct identification of gene function. In addition, it is a complex task to produce mice tailored to the needs of lung physiologists, because the lung is a very complex organ with numerous resident as well as motile cell types. To supplement this technical resource, other animal models, especially those, which have a far lesser complexity, would greatly facilitate research in this area.

Drosophila as a Model in Biomedical Research

Developments in recent years revealed that biomedical research critically depends on suitable animal models to understand the molecular basis underlying the pathogenesis of human diseases and to provide systems for developing and testing new therapies. Despite the supremacy of murine models, other model organisms are able to provide new and relevant information. Model organisms have to fulfil some criteria including sequenced genomes, short life cycles, similarities with human genes/proteins, and the ease of genetic manipulation, which is of prime importance. Among the limited number of well-established and generally accepted model organisms (yeast, *C. elegans*, *Drosophila*, zebra fish, and mice), the fruit fly *Drosophila* is the *primus inter pares*. It is the oldest model organism and was introduced almost a century ago by Thomas Hunt Morgan. Sequencing and analysis of its genome revealed a completely unforeseen degree of similarity with our own genome. More than 60% of all human disease genes have homologous counterparts in the fly (Fortini et al. 2000), which led to the development of a special database listing all these candidate genes (Chien et al. 2002). Since then, *Drosophila* models for a large number of different human diseases have been developed (Bier 2005). Among the first ones were those focusing on neurodegenerative diseases (Feany and Bender 2000). In addition to this Parkinson model, very informative models of Huntington's and Alzheimer's disease have been established (Chan and Bonini 2000), which triggered a great number of follow-up studies, finally leading to new therapeutic ideas. The plethora of models based on the fruit fly comprises those for the analysis of cardiac diseases (Wolf et al. 2006) or diabetes (Baker and Thummel 2007) to mention only two out of a huge number. For the unprejudiced reader it may be hard to understand why *Drosophila* should be that well suited for the study of human disease as the anatomical organization of the fruit fly is much simpler than that of mice. The major question is what makes *Drosophila* so special? *Drosophila* is simple enough to be a highly facile model but at the same time, major organs, physiological processes, and behaviours are similar to those found in higher mammals. This makes comparison between humans and flies much more relevant than comparisons between humans and worms or between humans and yeast. Together with the vast amount of knowledge that has been accumulated during the last century and the availability of countless technical opportunities to manipulate the fly, *Drosophila* is a model organism beyond comparison.

In the last couple of years, an incredibly rich source of tools has been developed allowing the study of the functional significance of genes of interest. Numerous deletions and far more than 12,000 transposable element insertions are readily available for creating genomic mutations. Transgenesis is very easy and there are a number of corresponding vectors available and some of them have been developed that capitalize on *P* element, phiC31 and recombineering technologies (Bellen et al. 2004; Venken and Bellen 2007; Venken et al. 2009). The most widely used and presumably the most versatile tool is the binary GAL4/UAS-system invented more than 15 years ago (Brand and Perrimon 1993). It relies on the yeast transcription factor GAL4 and the corresponding DNA binding region UAS that constitutes an extremely useful construction kit. Gal4/UAS systems conveniently allow overexpression or silencing of genes of interest wherever the researcher requires it. Both components (Gal4 and UAS) are segregated into different fly lines, called driver and effector lines, respectively. Effector lines (UAS-lines) determine what is expressed and driver lines (Gal4-lines) determine the stage of development and location of expression. Countless different driver (Gal4) lines are available, enabling the researcher to address almost every fly tissue separately (Fig. 5.1). Similarly, a great number of effector lines are available or can be generated easily by producing corresponding flies carrying a construct in which, for example, overexpression of a certain gene can be achieved because it is under transcriptional control of particular UAS elements. Crossing these two types of lines activates this system in the entire F1 generation (Fig. 5.1). This simple construction kit allows design of numerous different experiments with a small number of fly stocks, simply by means of combination. As the parental strains are usually not impaired, researchers can induce silencing of various genes of interest in one organ or overexpression of another gene in a different set of organs simply by crossing the corresponding lines. To complement this armamentarium, Barry Dickson's group completed the great endeavor of producing RNAi lines for almost all annotated *Drosophila* genes (Dietzl et al. 2007), and made them accessible to the *Drosophila* community. Combination of these publicly available lines with the GAL/UAS system allows efficient gene silencing in most parts of the organism or in an organ-specific fashion, depending on the GAL4 driver that is utilized. This unique resource enables even large-scale studies at reasonable prices and in a reasonable period of time.

A recent very useful refinement of this method allows additional temporal control of expression. The so-called TARGET system (McGuire et al. 2003) includes a temperature-sensitive repressor of GAL4, referred to as GAL80ts. It represses expression under restrictive conditions (19°C). A Shift to the permissive temperature (29°C) inactivates the temperature-sensitive repressor, thus releasing GAL4 from its inhibition, which leads to expression of the target gene. Usually Gal80ts is expressed under control of the ubiquitously active tubulin promotor, thus enabling effective inhibition in all organs at almost all time points.

More complex experiments became possible by the addition of alternative binary expression systems, namely the lexA-based binary expression system and the so-called Q-system (Potter et al. 2010). Many other powerful techniques exist that maximize the abilities of what can be studied in the fly. It is impossible to summarize all of them, but the reader is referred to additional reviews for more detailed discussion of recent developments in the field of *Drosophila* genetics (Bellen et al. 2004; Venken and Bellen 2007; Venken et al. 2009).

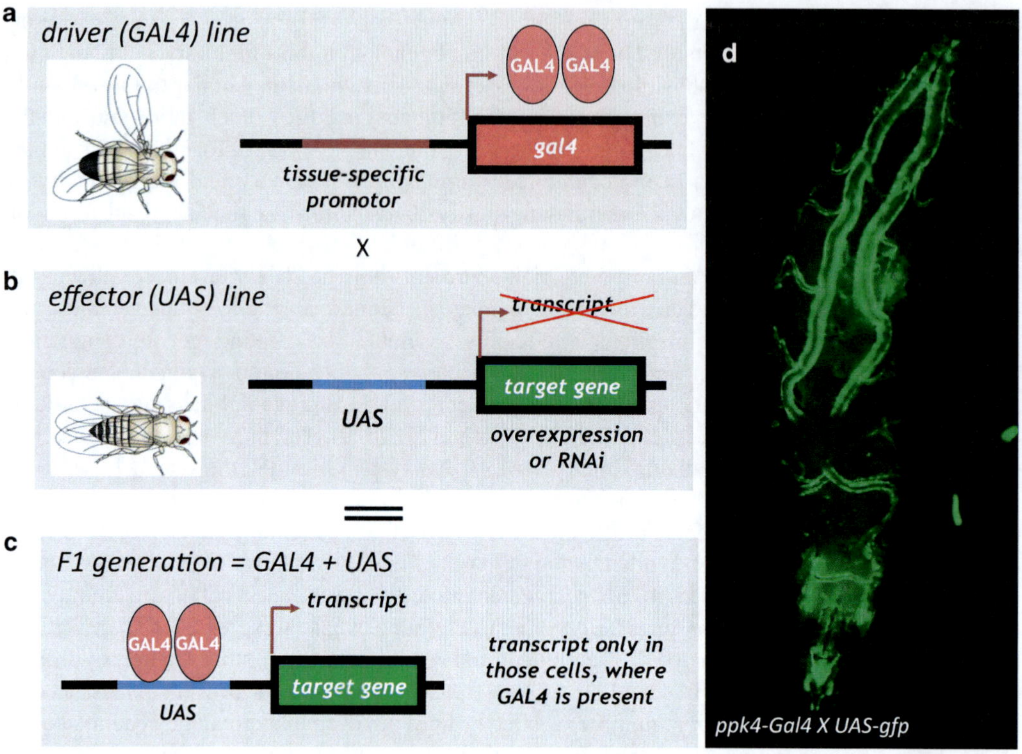

Fig. 5.1 Binary expression control systems allow targeted manipulation in the fly. A number of different binary expression systems are available for the fly. Most important is the Gal4/UAS system derived from yeast. Here, a so-called driver or Gal4 line determines the temporal and spatial expression pattern via expression of the yeast transcription factor Gal4 (**a**). A second type of line, the so-called effector or UAS-line has the target gene under transcriptional control of the UAS enhancer, which is also borrowed from yeast. This one is activated by Gal4 only (**b**). In the F1 generation, both factors (Gal4) and the UAS enhancer come together to direct expression of the target gene in a manner controlled by the expression pattern of the Gal4 line (**c**). Using an airway epithelium specific driver line (ppk4-Gal4) and gfp as the effector, we could label the airways epithelium specifically (**d**)

The Airway System of the Fly

Regarding asthma, the airway epithelium is obviously the most important target. Flies do not have lungs, but a tracheal system that supplies the entire organism with oxygen. Although lungs and tracheal systems are not homologous, a huge number of similarities can be observed at the level of the most important cell type comprising the corresponding organs, namely the airway epithelial cell. In *Drosophila*, the airway epithelium is a single cell layer surrounding the central airway. The airway system, or the tracheal system is made of interconnected tubes starting from major branches to secondary ones and ending in terminal branches supplying almost every cell of the fly (Fig. 5.2). Contact to the outside world is made through spiracles, which ensure an optimal oxygen flow (Affolter et al. 2003; Ghabrial et al. 2003; Ruehle 1932; Whitten 1957).

Important to note is that the airway system of the fly is made of epithelial cells only, meaning that no other cell types confound signals from these cells during experimental analysis.

The airway system is able to react towards an infection with an epithelial immune response, carried out by the immune competent epithelial cells. All airway epithelial cells have this capability, which is mainly characterized by the expression of antimicrobial peptide genes (Wagner et al. 2008).

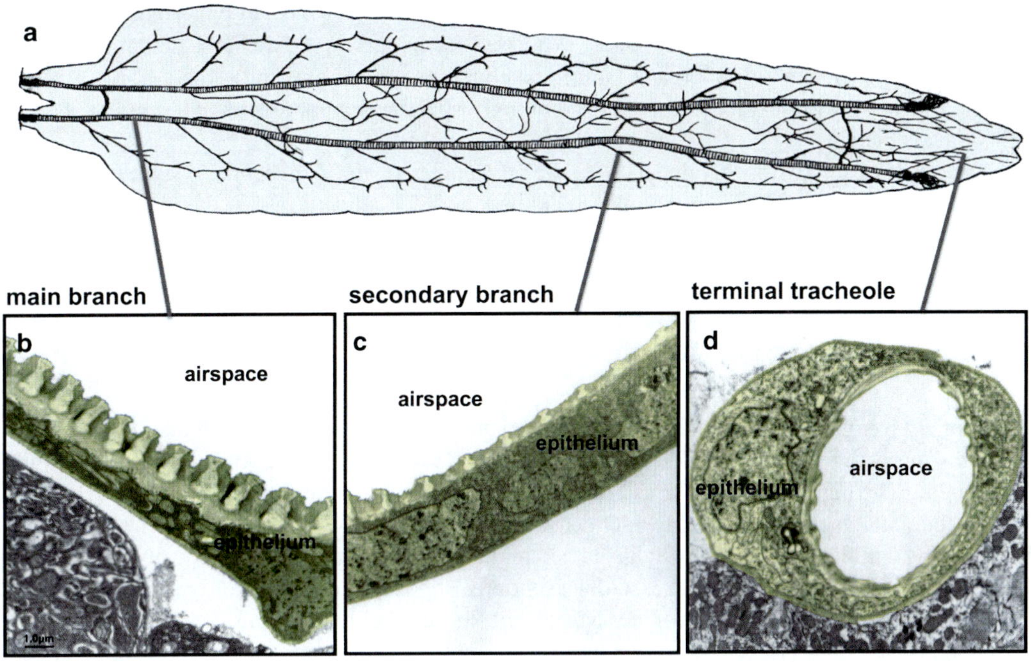

Fig. 5.2 The airway system of the fly has a very simple organization. It is composed (here a schematic drawing of a larva; **a**) of interconnected tubes that are organized in a hierarchical order. These tubes contain a central air filled lumen that is enclosed by a single layered epithelium. This organization is seen throughout the entire system ranging form the primary branches (**b**), including the secondary ones (**c**) and ending in the terminal tracheoles that supply all cells of the body with oxygen (**d**)

A detailed analysis of the signalling pathways present in the fly's airway epithelium and the pattern recognition molecules revealed that only one pathway converging on NF-kB activation, the IMD pathway, is operational in these tissues. In addition, the majority of pattern recognition molecules that the fly has to its disposal are present and functional there. It should be mentioned that other signalling pathways with relevance for immune reactions, including the JAK/STAT- and JNK-pathway are also functional in these cells (Wagner et al. 2008). All aspects of adaptive immunity are absent in the fly, making an unveiled view to the innate immune system possible, that shares most structural and architectural features with our own. Other physiological similarities include the way liquid clearance is achieved, just to mention one among a large number of examples.

Manipulation of the Airway Epithelium of the Fly

The airway system of the fly has been one of the prime models to study basic mechanisms underlying angiogenesis and comparable developmental traits. A huge number of studies have been performed with the goal to elucidate the mechanisms underlying tissue formation and development of complex structures. As these studies have almost completely been restricted to embryogenesis, the airway system of larvae and adults is much less well understood. Nevertheless, these studies have yielded a great number of highly relevant tools with which the airway system can be manipulated. Most important is obviously the Gal4/UAS system that allows all kinds of manipulation. To gain access to this enormously rich armament of possibilities, Gal4 lines that exclusively address the airway epithelium are mandatory. In larvae, we have some of these lines to our hand, namely btl-Gal4 and ppk4-Gal4. Both are exclusively expressed in the airway epithelium, without any confounding signal in any other organ. Unfortunately, this very promising situation is currently restricted to work with larvae, as the correspondingly specific drivers for the adult airway system are not available yet.

Thus, most of this work that depends on the advantages of the *Drosophila* system is currently performed with larvae, which on the other hand has the great advantage that the airway system is easily accessible to all kinds of microscopic analyses even in the living animal.

Drosophila in Asthma Research

The fruit fly has gained acceptance as a model organism to study aspects of human diseases, but asthma is surely one disease that doesn't initially spring to mind. Flies lack most characteristics that are thought to be highly relevant for asthma, such as adaptive immunity including IgEs and T-cells. In addition, outcomes of the disease such as wheezing and airway hyper-responsiveness have not been observed in flies. However, insights gained in the last few years have dramatically changed the way asthma is viewed – from a disease that is strictly dependent on the adaptive immune system only, to one where the innate immune system and the epithelial cells are of prime importance (Hammad et al. 2009; Holgate 2007). Apparently, epithelial cells orchestrate most adverse effects associated with this disease in a yet not well-understood way. Another central aspect relevant for asthma development is the innate immune system and especially the innate immune system within the airway epithelial cells. It became clear that those innate immune signaling pathways that converge on activation of NF-kB factors are of central importance for various aspects of disease development and progression (Broide et al. 2005; Pantano et al. 2008). Both of these newly identified hallmarks of asthma pathogenesis, the involvement of the innate immune system and NF-kB, brought the fly back into focus as a relevant model, simply because their contribution to disease development can be studied easily in the fly (Lemaitre and Hoffmann 2007; Roeder et al. 2009). Parallel to these developments, another set of asthma-related research also significantly changed the way the disease is viewed. Genetic research has revealed a number of asthma susceptibility genes (Vercelli 2008). In addition, with the development of genome-wide association studies, the roles of many more candidate genes in asthma will be uncovered. Surprisingly, most of these asthma susceptibility genes have nothing to do with adaptive immunity. Instead they are associated with innate immunity and with ways of ensuring epithelial integrity. Susceptibility genes that can be classified into either of these categories (innate immunity and epithelial integrity) comprise approximately 60% of the total. Except for those associated with adaptive immunity, we and others found unequivocal orthologues for all the rest (Roeder et al. 2009). Most asthma-related genes are expressed in the airway epithelium of the fly. Even more interestingly, some of them are regulated upon airway infection of the *Drosophila* airway epithelium, pointing to an important role in airway immunity and development of asthma-like phenotypes in the fruit fly. Indeed, what was previously thought to be a major disadvantage of flies for asthma research, the absence of adaptive immunity, may actually be a significant advantage for the researcher, because it allows an exclusive view on the role of innate immunity under different physiological conditions. In mice, it is a very complex task to distinguish between responses of the adaptive and the innate immune system, because they are highly interconnected.

Two different complementary experimental approaches are used in asthma-related *Drosophila* research. The first one is based on the observation that manipulation of innate immune systems within the epithelial cell is sufficient to induce structural changes very similar to those observable in chronic asthma patients or in murine models of the disease. The other approach depends on use of the *Drosophila* model for elucidation of the physiological as well as the pathophysiological significance of asthma susceptibility genes.

The first route of studies was initiated earlier in our lab and is based on the observation that the epithelial immune system of the fly's airway epithelium responds to an infection with bacteria or fungi not only with the expression of anti-microbial peptide genes, but additionally shows other reaction phenotypes that are relevant for asthma development. These phenotypical changes occur after prolonged and very strong activation of the epithelial immune response. Those parts of the tracheal system that react most strongly to an airway infection show significant airway remodelling. The remodelling of airways is one hallmark of chronic asthma. In the fly, the most conspicuous sign of remodeling is a several fold increase in epithelial thickness. This is accompanied by the production of new airway epithelial cells, something that usually does not occur during normal larval development in the fly. Taken together, these cells become meta- and hyperplastic, thus showing hallmarks of airway remodeling upon a prolonged strong activation of this (very special) type of immune system (Wagner et al. 2009). One major reason for this unprecedented type of reaction may be that these cells forcefully try to obviate apoptosis. Following infection highly potent apoptosis-inducing genes are activated, including the autophagy-specific kinase 1 gene (Atg1), whose overexpression itself is sufficient to induce apoptosis. Activation and overexpression of dFoxo appears to be a way to enable survival under these hostile conditions, which in turn, may be the major reason for the observed phenotypic variations (Wagner et al. 2009).

The second experimental route is a straightforward one, taking advantage of the entire toolkit available in *Drosophila* genetics. As already mentioned, more asthma susceptibility genes are involved in innate immune responses or in maintaining epithelial cell integrity than in processes associated with adaptive immunity. Innate immune signalling within these epithelial cells is necessary and sufficient for experimental disease induction, either in mice or flies. The central position of epithelial cells for orchestration of for example airway remodeling has been emphasized in recent physiological studies.

To show that *Drosophila*, with all its advantages, can be used as a model for asthma research, we analyzed the canonical set of asthma susceptibility genes and tried to find homologues in the fly (Roeder et al. 2009). Some of these genes, associated with adaptive immunity, have no counterparts in the fly. These genes include: IL-4, IL-10, IL-13, IL-18, IL-4R, CC16, CTLA4, HLA-DBP1, HLADQB1, HLA-DRB1, LTA, LTC, and FCERIB. For all other candidates, we found orthologues in the fly genome. This is summarized in Table 5.1. Most of them are indeed expressed in the fly's airway system and some of them (the majority) are regulated upon infection or ectopic manipulation of the IMD-signalling pathway. For most of these *Drosophila* counterparts, genetic tools are available comprising conventional knock-out strains, those that allow tissue-specific gene silencing, and additionally in some cases, those that allow tissue-specific overexpression (Table 5.1). Obviously, the methods to manipulate the airway system of the fly or orthologues of asthma susceptibility genes within this tissue are numerous, but what about simple read-outs? Identification of disturbances within the tracheal system can be monitored using two very simple but nevertheless informative approaches. The transparency of the larvae allows a visual evaluation of the tracheal structure such that functional impairments, which are often caused by liquid-filled parts of the tracheal lumen, are easily identified. A physiological test of the operational capability of the tracheal system takes advantage of the normal behaviour of the larvae. Healthy animals are usually buried into their food, but upon functional oxygen deprivation, they move to the surface. This is also observed in larvae that show impairments of the tracheal system. This behaviour can be quantified (time to leave food or percentage of larvae showing this behaviour). The reason for choosing a seemingly simple model organism is that one hopes to be confronted with a significantly reduced complexity of the genomic situation, meaning much lesser redundancy (number of genes of a certain gene family) and much lesser plasticity (meaning that genes can take the role of a deleted one). Both aspects are obviously on the side of the fly. Usually, gene families that

Table 5.1 Asthma susceptibility genes and their homologues in *Drosophila*

Asthma gene	Drosophila homologue	Expressed in airways	Regulated or relevant in airways	Genomic resources
ACE	ANCE-family	+	ance-2	RNAi
ADAM33	Neu3(CG7649), mmd(CG42252)	?,?		all:KO,RNAi
ADRB2	Oct2ß-like	+	Oct2ß-like	KO,RNAi
CCL11,CCL5	upd1-3	+,+,?	upd-2	KO,RNAi
CD14	CG5195	?		RNAi
CMA1	trypsin29F,snk	?,+		RNAi;KO,RNAi
DPP10	ome,CG17684	?,+		both:KO,RNAi
FLG	CG1433 Atu	?		KO,RNAi
GPRA	CcapR,AlCR2, TakR86C	?,+,+	AlCR2	all:KO,RNAi
GSTM1	GST-D family	+	D1,D2,D3,D9	all:KO,RNAi
GSTT1	GST-E family	+	E1,E5,E7	all:KO,RNAi
GSTP1	GST-S1	?		KO,RNAi
MMP9	DmMMP1	+	DmMMP1	KO,RNAi,OE
NAT2	aaNAT-1	?		KO,RNAi
NOS	dNOS	+		KO,RNAi,OE
ORMDL3	dORMDL	+		RNAi
STAT6	STAT92	+	STAT92	KO,RNAi,OE
SPINK5	Serpins	+	spn77Ba	KO,RNAi
TBXA2R	CG7497,OAMB	?,+		KO,RNAi; KO,RNAi,OE
TGFb1	activin, maverick, gbp	?,?,?		RNAi, -; RNAi
TLR2, 4, 6, 10	Toll receptors	+	Toll, 18-w, Toll-7, -8	all: KO, RNAi
TNF	Eiger	?		RNAi

Listed are the orthologues of canonical asthma susceptibility genes as well as the presence and regulation in the airway epithelium. In addition, the available genomic resources have been extracted from the VDRC (Dietzl et al. 2007) or flybase (www.flybase.org).

have numerous members in mammals are comprised of fewer or sometimes only one member in the fly, which is exemplified by the matrix metalloproteinases (MMPs). Humans have 23 MMPs with different but partially overlapping physiological roles, whereas in *Drosophila* only two MMPs are present and only one of them is expressed in the airway epithelium (Page-McCaw et al. 2003).

Potential Roles of Asthma Susceptibility Genes in *Drosophila*

To show the potential of the Drosophila system, we would like to discuss our current state of knowledge regarding some of the homologues of asthma susceptibility genes. Among the matrix metalloproteinases (MMP), at least two, namely MMP9 and MMP7, are considered as highly relevant for asthma (Table 5.1). Using the sole airway epithelium relevant homologue found in *Drosophila* (DmMMP1), their role could be studied in order to tease apart its function independent of the other 21 MMPs (Goswami et al. 2009). Nevertheless, we have almost no experimentally based information regarding its direct role in asthma development and progression. In *Drosophila*, DmMMP1 has been shown to be important for tracheal development (Page-McCaw et al. 2003).

DmMMP1 and MMP9 are both secreted and they accept collagen as a substrate, pointing to similar functions. This suggests that a closer look at the physiological role of DmMMP1 in airway biology using the methodological toolkit available in *Drosophila* will allow us a better understanding of MMP9's and MMP7's contribution to asthma pathology.

Another important group of asthma susceptibility genes are the signal transducer and activators of transcription (STATs). In the fly, the complexity of JAK/STAT signalling is reduced to its minimum. In mammals, various families of ligands, receptors, JAK kinases and STATs are present, whereas in *Drosophila* only one receptor, one JAK kinase, and one STAT constitute this pathway (Arbouzova and Zeidler 2006; Shuai and Liu 2003). Since both, *STAT6* and *STAT3* are believed to play important roles in asthma pathogenesis, the presence of only one STAT homologue in the fly provides a unique opportunity to study the physiological significance of this factor in the airways. All components of the pathway, including the ligand *unpaired* (*upd*), are expressed in airway epithelial cells. In addition, we have shown that STAT-dependent signalling is operational in defined parts of the airways, even without pathogen contact (Wagner et al. 2008). A more detailed analysis of JAK/STAT signalling in the airway epithelium of the fly has therefore great potential to provide new information about its role in health and disease of the airways.

Another class of genes that appears to be relevant for asthma pathogenesis also has some members in the fly. These are the serpins (serine protease inhibitors) and one member, namely *SPINK5*, is a well-established asthma susceptibility gene. The copious number of genes with seemingly very similar functional roles complicates understanding the physiological and pathophysiological role of serpins. A functional equivalent in *Drosophila* is the serpin *Spn77Ba*. This serpin is present in the fly's airway epithelium, where it plays an important role as a gatekeeper of the epithelial innate immune system. Its disruption induces strong activation of the epithelial but also of the systemic immune system. The latter response appears to be mediated by an unknown pathway, finally leading to release of anti-microbial peptides from the fat body, which is triggered by activation of the Toll pathway (Tang et al. 2008). This implies that serpins act as controllers of epithelial immunity, ensuring that immune-relevant proteases are inhibited in times of normal activities. Their deregulation may induce prolonged inflammation and thereby contribute to asthma development and chronification.

Completely different types of molecules that are highly relevant in asthma are G-protein coupled receptors as major transducers of information. Adrenergic receptors are of special interest, because they are targeted by a huge number of asthma-related pharmaceuticals. With respect to this, the most important may be the fly's orthologue of the adrenergic receptor, called *oct2β*, which is expressed in the airway epithelium. It shares not only structural similarities but appears to mediate similar effects in the airways via similar second-messenger systems. The methods available to directly manipulate second-messenger systems in the airway epithelia can be used to study the effects of long-term medication with adrenergics, an aspect that is not understood at all.

The last group of genes that should be discussed are the glutathione-S transferases (GST), which is a very copious gene family in our genome. Some of them are asthma susceptibility genes. They are believed to function as general enzymatic antioxidants that detoxify xenobiotics and facilitate normal lung function when their expression is increased in response to different types of stress. Surprisingly, we know almost nothing about their physiological role in the normal lung, and we also have limited experimental information about their significance in asthma development. However, all major GSTs known to be relevant for asthma pathogenesis have orthologues in the fly. All these orthologues are expressed in the airways, allowing their functional analysis in *Drosophila*. The full usefulness of the *Drosophila* model may emerge in combination with the most powerful genetic tools of today: genome-wide association studies (GWAS) and second generation, whole genome deep sequencing. These new techniques promise that discoveries of novel disease susceptibility loci, genes and disease-associated gene variants will reach unprecedented speed. As most of the genes located in susceptibility loci will presumably be of unknown function and their involvement in asthma will hot have been documented, a quick but comprehensive test regarding their potential role in airway physiology and pathophysiology is mandatory. While genotyping and sequencing have made quantum leaps forward, functional genomics studies are still tedious, expensive, and time consuming. As such, there is the danger that we may overlook major parts of genome-wide association studies. Thus, translation of these results into basic asthma research and clinical applicability may be slowed down or hampered considerably at this stage.

Drosophila is thus a valuable tool that might enable us to categorize putative candidate genes for further downstream analyses in vertebrate models, including murine models. *Drosophila* may be used to dissect the fraction by which individual genes of a gene cluster contribute causally to an observed susceptibility signal. Even more importantly, it can be used to identify the relevance of a gene in a disease pathway or identify stimuli relevant for gene regulation. These potential applications are relevant for all complex diseases but especially for lung diseases such as asthma, because the mammalian organ of interest is difficult to access and resident as well as motile elements contribute to disease development and progression. Using *Drosophila* models, preliminary experiments with other species (especially with mice) may be reduced significantly, allowing a better focus for these time-consuming studies. Utilizing *Drosophila* in a systematic approach together with other models and tools seems most promising and may significantly reduce the turnaround time from genetic results into biologically meaningful data.

Taken together, *Drosophila* has the potential to develop into a complementary model in asthma research. The inherent strengths of this system can open a novel field in asthma research that might enable us to understand, at least in part, the role of important asthma susceptibility genes in detail.

Acknowledgements Research in our group was sponsored by the German Research Foundation (DFG) as parts of the SFB Transregio-22 (Teilprojekt A7) and the Cluster Inflammation@interfaces.

References

Affolter M, Bellusci S, Itoh N, Shilo B, Thiery JP, Werb Z (2003) Tube or not tube: remodeling epithelial tissues by branching morphogenesis. Dev Cell 4(1):11–18

Arbouzova NI, Zeidler MP (2006) JAK/STAT signalling in Drosophila: insights into conserved regulatory and cellular functions. Development 133(14):2605–2616

Baker KD, Thummel CS (2007) Diabetic larvae and obese flies-emerging studies of metabolism in Drosophila. Cell Metab 6(4):257–266

Bellen HJ, Levis RW, Liao G, He Y, Carlson JW, Tsang G, Evans-Holm M, Hiesinger PR, Schulze KL, Rubin GM, Hoskins RA, Spradling AC (2004) The BDGP gene disruption project: single transposon insertions associated with 40% of Drosophila genes. Genetics 167(2):761–781

Bier E (2005) Drosophila, the golden bug, emerges as a tool for human genetics. Nat Rev Genet 6(1):9–23

Brand AH, Perrimon N (1993) Targeted gene expression as a means of altering cell fates and generating dominant phenotypes. Development 118(2):401–415

Broide DH, Lawrence T, Doherty T, Cho JY, Miller M, McElwain K, McElwain S, Karin M (2005) Allergen-induced peribronchial fibrosis and mucus production mediated by IkappaB kinase beta-dependent genes in airway epithelium. Proc Natl Acad Sci USA 102(49):17723–17728

Chan HY, Bonini NM (2000) Drosophila models of human neurodegenerative disease. Cell Death Differ 7(11):1075–1080

Chien S, Reiter LT, Bier E, Gribskov M (2002) Homophila: human disease gene cognates in Drosophila. Nucleic Acids Res 30(1):149–151

Dietzl G, Chen D, Schnorrer F, Su KC, Barinova Y, Fellner M, Gasser B, Kinsey K, Oppel S, Scheiblauer S, Couto A, Marra V, Keleman K, Dickson BJ (2007) A genome-wide transgenic RNAi library for conditional gene inactivation in Drosophila. Nature 448(7150):151–156

Feany MB, Bender WW (2000) A Drosophila model of Parkinson's disease. Nature 404(6776):394–398

Finkelman FD, Wills-Karp M (2008) Usefulness and optimization of mouse models of allergic airway disease. J Allergy Clin Immunol 121(3):603–606

Fortini ME, Skupski MP, Boguski MS, Hariharan IK (2000) A survey of human disease gene counterparts in the Drosophila genome. J Cell Biol 150(2):F23–F30

Ghabrial A, Luschnig S, Metzstein MM, Krasnow MA (2003) Branching morphogenesis of the Drosophila tracheal system. Annu Rev Cell Dev Biol 19:623–647

Goswami S, Angkasekwinai P, Shan M, Greenlee KJ, Barranco WT, Polikepahad S, Seryshev A, Song LZ, Redding D, Singh B, Sur S, Woodruff P, Dong C, Corry DB, Kheradmand F (2009) Divergent functions for airway epithelial matrix metalloproteinase 7 and retinoic acid in experimental asthma. Nat Immunol 10(5):496–503

Hammad H, Chieppa M, Perros F, Willart MA, Germain RN, Lambrecht BN (2009) House dust mite allergen induces asthma via Toll-like receptor 4 triggering of airway structural cells. Nat Med 15(4):410–416

Holgate ST (2007) The epithelium takes centre stage in asthma and atopic dermatitis. Trends Immunol 28(6):248–251

Holgate ST, Polosa R (2008) Treatment strategies for allergy and asthma. Nat Rev Immunol 8(3):218–230

Holloway JW, Yang IA, Holgate ST (2008) Interpatient variability in rates of asthma progression: can genetics provide an answer? J Allergy Clin Immunol 121(3):573–579

Lemaitre B, Hoffmann J (2007) The host defense of Drosophila melanogaster. Annu Rev Immunol 25:697–743

McGuire SE, Le PT, Osborn AJ, Matsumoto K, Davis RL (2003) Spatiotemporal rescue of memory dysfunction in *Drosophila*. Science 302(5651):1765–1768

Page-McCaw A, Serano J, Sante JM, Rubin GM (2003) Drosophila matrix metalloproteinases are required for tissue remodeling, but not embryonic development. Dev Cell 4(1):95–106

Pantano C, Ather JL, Alcorn JF, Poynter ME, Brown AL, Guala AS, Beuschel SL, Allen GB, Whittaker LA, Bevelander M, Irvin CG, Janssen-Heininger YM (2008) Nuclear factor-kappaB activation in airway epithelium induces inflammation and hyperresponsiveness. Am J Respir Crit Care Med 177(9):959–969

Potter CJ, Tasic B, Russler EV, Liang L, Luo L (2010) The Q system: a repressible binary system for transgene expression, lineage tracing, and mosaic analysis. Cell 141:536–548

Roeder T, Isermann K, Kabesch M (2009) Drosophila in asthma research. Am J Respir Crit Care Med 179(11):979–983

Ruehle H (1932) Das larvale Tracheensystem von Drosophila melanogaster Meigen und seine Variabilität. Z Wiss Zool 141:159–245

Shuai K, Liu B (2003) Regulation of JAK-STAT signalling in the immune system. Nat Rev Immunol 3(11):900–911

Tang H, Kambris Z, Lemaitre B, Hashimoto C (2008) A serpin that regulates immune melanization in the respiratory system of Drosophila. Dev Cell 15(4):617–626

Venken KJ, Bellen HJ (2007) Transgenesis upgrades for Drosophila melanogaster. Development 134(20):3571–3584

Venken KJ, Carlson JW, Schulze KL, Pan H, He Y, Spokony R, Wan KH, Koriabine M, de Jong PJ, White KP, Bellen HJ, Hoskins RA (2009) Versatile P[acman] BAC libraries for transgenesis studies in Drosophila melanogaster. Nat Methods 6(6):431–434

Vercelli D (2008) Discovering susceptibility genes for asthma and allergy. Nat Rev Immunol 8(3):169–182

Wagner C, Isermann K, Fehrenbach H, Roeder T (2008) Molecular architecture of the fruit fly's airway epithelial immune system. BMC Genomics 9:446

Wagner C, Isermann K, Roeder T (2009) Infection induces a survival program and local remodeling in the airway epithelium of the fly. FASEB J 23(7):2045–2054

Whitten J (1957) The post-embryonic development of the tracheal system in Drosophila melanogaster. Q J Microsc Sci 98:123–150

Wolf MJ, Amrein H, Izatt JA, Choma MA, Reedy MC, Rockman HA (2006) Drosophila as a model for the identification of genes causing adult human heart disease. Proc Natl Acad Sci USA 103:1394–1399

Chapter 6
Elucidating the In Vivo Targets of *Photorhabdus* Toxins in Real-Time Using *Drosophila* Embryos

Isabella Vlisidou, Nicholas Waterfield, and Will Wood

Abstract The outcome of any bacterial infection, whether it is clearance of the infecting pathogen, establishment of a persistent infection, or even death of the host, is as dependent on the host as on the pathogen (Finlay and Falkow 1989). To infect a susceptible host bacterial pathogens express virulence factors, which alter host cell physiology and allow the pathogen to establish a nutrient-rich niche for growth and avoid clearance by the host immune response. However survival within the host often results in tissue damage, which to some cases accounts for the disease-specific pathology. For many bacterial pathogens the principal determinants of virulence and elicitors of host tissue damage are soluble exotoxins, which allow bacteria to penetrate into deeper tissue or pass through a host epithelial or endothelial barrier. Therefore, exploring the complex interplay between host tissue and bacterial toxins can help us to understand infectious disease and define the contributions of the host immune system to bacterial virulence. In this chapter, we describe a new model, the *Drosophila* embryo, for addressing a fundamental issue in bacterial pathogenesis, the elucidation of the in vivo targets of bacterial toxins and the monitoring of the first moments of the infection process in real-time. To develop this model, we used the insect and emerging human pathogen *Photorhabdus asymbiotica* and more specifically we characterised the initial cross-talk between the secreted cytotoxin Mcf1 and the embryonic hemocytes. Mcf1 is a potent cytotoxin which has been detected in all *Photorhabdus* strains isolated so far, which can rapidly kill insects upon injection. Despite several in vitro tissue culture studies, the biology of Mcf1 in vivo is not well understood. Furthermore, despite the identification of many *Photorhabdus* toxins using recombinant expression in *E. coli* (Waterfield et al. 2008), very few studies address the molecular mechanism of action of these toxins in relation to specific immune responses in vivo in the insect model.

Introduction

Although *Drosophila* has been widely used to study immune responses against bacterial species pathogenic to humans (Shirasu-Hiza and Schneider 2007; Pielage et al. 2008; Kim et al. 2008; Frandsen et al. 2008; Brandt et al. 2004; Blow et al. 2005; Fleming et al. 2006), the battle between insect pathogens, or pathogens capable of infecting both insects and humans, and *Drosophila*

I. Vlisidou (✉) • N. Waterfield • W. Wood
Department of Biology and Biochemistry, University of Bath, BA2 7AY Bath, UK
e-mail: iv203@bath.ac.uk

E. Mylonakis et al. (eds.), *Recent Advances on Model Hosts*,
Advances in Experimental Medicine and Biology 710, DOI 10.1007/978-1-4419-5638-5_6,

immune cells has not been studied in detail. *Photorhabdus* is a Gram-negative member of the family of *Enterobacteriaceae* and the only bioluminescent terrestrial bacterial species (Forst et al. 1997; Waterfield et al. 2009). All members of this genus are highly motile, facultative anaerobic rods (Peel et al. 1999), which grow at 28°C (for all species) and from 37°C to 42°C (for clinical species only). The genus comprises three species: *P. temperata* and *P. luminescens*, both of which are strict insect pathogens; and *P. asymbiotica*, which are also capable of causing clinical infections to healthy humans (Farmer et al. 1989; Gerrard et al. 2003, 2004; Fischer-Le Saux et al. 1999). All species live in a symbiotic association with an entomopathogenic *Heterorhabditis* nematode and they have never been isolated directly from the environment (Forst et al. 1997). The nematode transmission cycle of *Photorhabdus* involves a complex process of cell-specific invasion of hermaphrodite adults and offspring nematodes (Ciche et al. 2008; Waterfield et al. 2009). The cycle starts with the developmentally arrested and highly stress resistant specialised larval stage of the nematode called the infective juvenile (IJ). This diapaused juvenile (a Daurer equivalent) hunts in the soil for insect prey which they penetrate to gain access to the open blood-system (hemocoel). Once in the blood, the IJs "recover" and regurgitate bacteria directly into thehemocoel (Ciche and Ensign 2003). Once in the hemocoel, the bacteria set up a lethal septicaemia, replicating (Daborn et al. 2002; Silva et al. 2002). They secrete a complex cocktail of toxins and proteases that rapidly kill the insect and allow bioconversion of its tissues into more bacteria, which in turn provide a food source for the developing nematodes. Later a sub-popultaion of ingested *Photorhabdus* begin to colonise the posterior part of the nematodes intestine, invading three rectal gland cells and remaining intracellular. The bacteria manipulate the development of the nematode, inducing endotokia matricida, in which the eggs are not laid, but hatch inside the body of the adult hermaphrodite. At this point, the bacteria are released from the rectal gland cells and proceed to colonise this new generation of L1 juveniles, trapped within the body cavity of the parent, via intracellular invasion of pharyngeal intestinal valve cells (Ciche et al. 2008). At this point the bacteria again appear to control nematode development, inducing the formation of IJs from the L1 worms, which are then free to leave the insect cadaver in search of new prey. Each IJ carries approximately 100 bacteria, a dose which is able to initiate a successful insect infection and suggests that the bacteria are adapted to be highly virulent and able to resist the immune systems of a very diverse range of insect orders, which may serve as prey.

The large Lepidopteran species *Manduca sexta* and *Galleria mellonela* are commonly used to understand *Photorhabdus* virulence (reviewed by Waterfield et al. 2009), although a recent study has also introduced the *Drosophila* larvae as an alternative model to decipher the tripartite interaction between the symbionts, the nematodes and the insect (Hallem et al. 2007). Following injection in *M. sexta Photorhabdus* rapidly multiplies in the hemolymph and colonises a specific niche between the extracellular matrix and the basal membrane of the midgut epithelium where it may be partially protected from the circulating hemocytes (Silva et al. 2002). Nevertheless, insect hosts are able to recognize *Photorhabdus* and the expression of several immune-regulated genes including hemolin, immunolectin-2 and the peptidoglycan recognition proteins in *M. sexta* (Eleftherianos et al. 2006a, b) have been shown to slow the otherwise rapid killing. Moreover antimicrobial peptides (AMP) including attacin, cecropin and moricin in *M. sexta* and metchnikowin, diptericin, drosomycin and attacin in *Drosophila* larvae can also be seen to be activated upon *Photorhabdus* infection (Eleftherianos et al. 2006b; Hallem et al. 2007). Despite the fact that insects are able to mount an immune response to *Photorhabdus*, it is not immunoprotective and the bacteria always prevail over this response (Eleftherianos et al. 2006a, b; Hallem et al. 2007). *Photorhabdus* also uses a type III secretion system to inhibit phagocytosis (Brugirard-Ricaud et al. 2005), although the effector(s) responsible for this activity are still largely unknown. Insects respond to pathogens and damaged tissue by activating the melanisation and encapsulation response mediated by phenoloxidase (PO) (Gillespie et al. 1997). *Photorhabdus* has evolved mechanisms to deal with the insect phenoloxidase pathway including the antibiotic molecule,

(E)-1-3-dihydroxy-2-(isopropyl)-5-(2-phenyethylenyl) benzene (known as ST), that inhibits phenoloxidase (Eleftherianos et al. 2007), and an inhibitor of the enzyme phospholipase A2 (PLA2) which is responsible for eicosanoid biosynthesis and haemocyte nodulation and prophenoloxidase activation (Kim et al. 2005).

A New Perspective: The *Drosophila* Embryo

It is evident that Lepidopteran insect models have contributed an enormous amount of important information related to the interplay between *Photorhabdus* and the innate immune system. Given their large size they have been ideal models to study interactions from a biochemical perspective. Nevertheless, the conclusions are derived from cohorts of experimental animals euthanized at multiple time points. These are essentially "snapshots" of in vivo responses at a relatively crude resolution considering the rate at which the underlying events take place. The monitoring of the initial stages of infection, particularly of a highly adapted entomopathogen such as *Photorhabdus*, is absolutely crucial in order to distinguish between pathogen-specific responses and responses related to gross changes in the insect due to hyper activation of the immune system which are evident at the late stages of infection. Due to the lack of real-time imaging techniques for the typically opaque larvae and adult stages of insects, activities such as the spread of a pathogen to an unexpected site of infection may be missed, unless such infected tissue was specifically harvested and analysed. Finally, conventional assays of pathogenesis typically require large numbers of animals to obtain statistically meaningful data at multiple time points. Given that, we used the *Drosophila* embryos as a complementary system to the conventional insect models of microbial pathogenesis (Vlisidou et al. 2009). *Drosophila* embryos and their development are extremely well documented and recent attention has focused on the role of the embryonic hemocytes in early embryonic development (Wood and Jacinto 2007). Embryonic hemocytes are highly motile cells that fulfil equivalent functions to mammalian neutrophils and macrophages. The cells follow specific migratory routes during embryogenesis dispersing from their point of origin in the procephalic mesoderm to eventually distribute themselves equally throughout the animal by late embryonic stages (Tepass et al. 1994; Mandal et al. 2004; Bataillé et al. 2005). As they migrate in the developing embryo they deposit extracellular matrix and engulf apoptotic cells, a process which is very important considering the extensive tissue reshaping taking place and the subsequent high number of apoptotic cells that need to be removed during embryogenesis (Wood and Jacinto 2007). They are also essential for the proper development of many key structures such as the gut and central nervous system (Olofsson and Page 2005).

Drosophila embryos combine the sophisticated genetics of the adult fly model with excellent microscopic tractability at a cellular level. Embryonic immune cells can be genetically modified to express fluorescent probes from tissue-specific promoters and their migrations can be followed using confocal timelapse microscopy. The function of phenotype-related genes can be tested by creating global mutants or by knocking down the expression of the gene of interest in specific tissues. Microbes or purified toxins can be introduced to the system relatively easily by microinjection and depending on the virulence capacity of the infectious agent studied; the embryos can survive mounted on a slide in gas-permeable oil undergoing normal development. The embryonic immune response has not been well studied to date. It has been assumed that early stage embryos would lack a complete immune response due to the perceived lack of interactions with the exterior world. Conversely, late embryos (stage 15 onwards) are preparing for the larval stage and express molecules which have been implicated in immunity at later stages of animal's life. As hemocytes found in adult and larval tissue originate from both embryonic and larval stages (Holz et al. 2003), it was particularly important for us to confirm that *Drosophila* embryos are able to raise immune response upon infection.

Using *Drosophila* Embryos to Study Bacterial Infection and Toxin-Induced Damage in Real-Time

Taking advantage of the embryo transparency and its amenability for live imaging, we managed to monitor the interaction of micro-injected, fluorescently-labelled non-pathogenic *E. coli* with embryonic haemocytes of stage 15 embryos. At this embryonic stage, hemocytes are arranged into three characteristic lines that run anterior to posterior along the ventral side of the embryo (Wood and Jacinto 2007). As early as 20 min after injection, haemocytes and bacteria co-localise on these developmental routes; suggesting that hemocytes are able to recognise and bind the bacteria, although it is still not clear whether the hemocytes migrate actively towards the bacteria. Interestingly, timelapse confocal microscopy revealed that haemocytes are able to engulf *E. coli* and optical sections collected at different focal planes show labelled bacteria within their cytoplasm. The fact that infected embryos are able to hatch into larvae and transform to adult flies indistinguishable from the wild-type embryos, suggests that embryonic haemocytes in collaboration with other innate defenses and/or nutrient restrictions within the embryo are capable of completely clearing the *E. coli* infection. The unfolding of these events nevertheless warrants further detailed investigation.

Most of the systemic responses in insects are activated by pattern recognition receptors (PRRs) that directly sense microbial elicitors. *Drosophila* immune-competent cells (fat body and hemocytes) have the potential to express more than 18,000 isoforms of the immunoglobulin (Ig)-superfamily receptor Down syndrome cell adhesion molecule Dscam (Wojtowicz et al. 2004, 2007) in the form of membrane-binding proteins that can serve as recognition receptors, while soluble isoforms of Dscam have also been detected in the hemolymph (Watson et al. 2005). RNAi-mediated gene silencing of Dscam in *Drosophila* 3rd instar larval haemocytes and *Anopheles gambiae* significantly decreases their efficiency in phagocytosing both Gram-positive and Gram-negative bacteria, suggesting that Dscam acts as either a recognition or signalling receptor during phagocytosis, perhaps analogous to vertebrate antibodies (Watson et al. 2005; Dong et al. 2006). Importantly, proteomics studies performed by Stuart et al. have demonstrated the presence of Dscam in the phagosome of S2 cells (Stuart et al. 2007). Despite the expression of Dscam on stage 15 embryonic hemocytes (W. Wood, unpublished results) and in contrast to the above observations *dscam* null embryos are able to recognise and crosslink bacteria on the surface of their hemocytes suggesting that Dscam is redundant for bacterial recognition in the embryo. However Dscam is a highly variable receptor therefore one can speculate that the diversification mechanisms that generate the massive expansion of receptors may not be fully functional at this specific stage of development.

Defence against pathogens partially relies on fast tissue infiltration by immunocompetent hemocytes that migrate from circulating hemolymph to a wound or to sites of infection, in order to deliver, *in situ*, an effective immune response. Surprisingly, infection of *Drosophila* embryos with *P. asymbiotica* produces an unusual phenotype whereby hemocytes rapidly lose their ability to migrate and apparently freeze within minutes of infection. As a result the haemocytes are not able to phagocytose, although their capacity for recognising and binding *Photorhabdus* on their surface remains unaffected. The phenotypes of hemocyte paralysis and inability to phagocytose can be recapitulated by injection of wild-type stage 15 embryos with *E. coli* constitutively expressing Mcf1 toxin or purified Mcf1 protein in a dose-dependent manner. The localisation of a toxin in infection models typically involves post mortem immunohistochemistry of tissue. However using the *Drosophila* embryo model, we were able to use for the first time fluorescently-labeled Mcf1 protein to directly localise the toxin to embryonic hemocytes in a living animal. The toxin was active only when it could be endocytosed by hemocytes as dynamin mutant embryos (Kitamoto 2001) were completely protected from the paralytic action of Mcf1. The embryo model also allowed us to address another question regarding the biology of this toxin; namely, does Mcf1 cause widespread epithelial tissue damage or is there specificity in its tissue target(s) in a whole animal system. To address this we analysed dorsal closure, a naturally occurring epithelial movement which requires the coordinated migration and fusion of two epithelial sheets to close the dorsal side of the embryo (Jacinto et al. 2002). The process requires the small GTPase Rac (Woolner et al. 2005), and a fully functional actin cytoskeleton (Jacinto et al. 2000). Interestingly, Mcf1 injected into embryos had no effect on dorsal closure and epithelial fronts migrated and fused at the same rate as the wild-type. Even though we cannot exclude the possibility that epithelial cells are less endocytic than hemocytes, we can still postulate that Mcf1 does not appear to affect all embryonic cell types in the dramatic fashion observed in hemocytes. This tissue specificity in the whole animal is in contrast to in vitro tissue culture studies which suggest Mcf1 is totally indiscriminate in cell types it can intoxicate. This observation with a single toxin species is very important in highlighting the relevance of host cell/tissue activities in influencing the outcome of an infection.

Another interesting point related to Mcf1 function is its ability to cause apoptosis on different epithelial cell lines by a novel BH3-domain-mediated effect (Dowling et al. 2004, 2007). The expression of the first pro-apoptotic markers occurs as early as 6 h post intoxification of epithelial cell lines with Mcf1 (Dowling et al. 2004). These data together with the rapid paralysis of hemocytes in the presence of Mcf1 suggests that this phenotype is independent, or upstream of, apoptosis given that the earliest pro-apoptotic indicators are observed 3 h following incubation with Mcf1 (Dowling et al. 2004). By comparing the hemocyte morphology in embryos expressing the pro-apoptotic Bcl-2 family member Bax in hemocytes using a combination of srpGAL4 and crqGAL4 drivers, with the hemocytes from wild-type embryos injected with Mcf1, we confirmed that the early freezing effect of Mcf1 is independent of apoptosis. Consistent with the long term apoptotic effect of Mcf1 the hemocytes of embryos injected with Mcf1 for 12 h appeared morphologically identical to those overexpressing Bax.

The rapid onset of the freezing phenotype suggested to us that Mcf1 destabilises cytoskeletal dynamics. Several bacterial toxins target Rho GTPases, which constitute molecular switches in several signalling processes and master regulators of the actin cytoskeleton (Etienne-Manneville and Hall 2002; Burridge and Wennerberg 2004). The small GTPase Rac has been previously implicated in phagocytosis and cell migration in mammals and has been shown to be essential for hemocyte migration within the embryo (Paladi and Tepass 2004; Stramer et al. 2005). Using both hemocyte dominant-negative (RacN17) or constitutively active (RacV12) versions of Rac we showed for first time that Mcf1-mediated paralysis requires Rac as both mutant embryos were completely resistant to Mcf1.

Conclusions and Future Perspectives

Here we have highlighted the power of the *Drosophila* embryo as a model for monitoring bacterial infection and localising toxin targets in a whole animal. We propose that this model has great potential in the future for characterising the trafficking of immune cells in response to intoxification, infection and wounding. Detailed studies of highly dynamic and motile immune cells demands the use of models that are not limited by time and allow us to monitor the earliest possible moments of the interaction between host and pathogen in the context of a living tissue. Although the *Drosophila* embryo as an infection model is in it's infancy, we envision that it could contribute to studies aiming to quantify changes in tissue architecture, permeability, or metabolism associated with infection and expect it to develop into an indispensable tool of basic and translational research in infectious disease. Finally this system also provides a unique model to understand the impact of infection upon normal embryonic development of a multi-cellular animal.

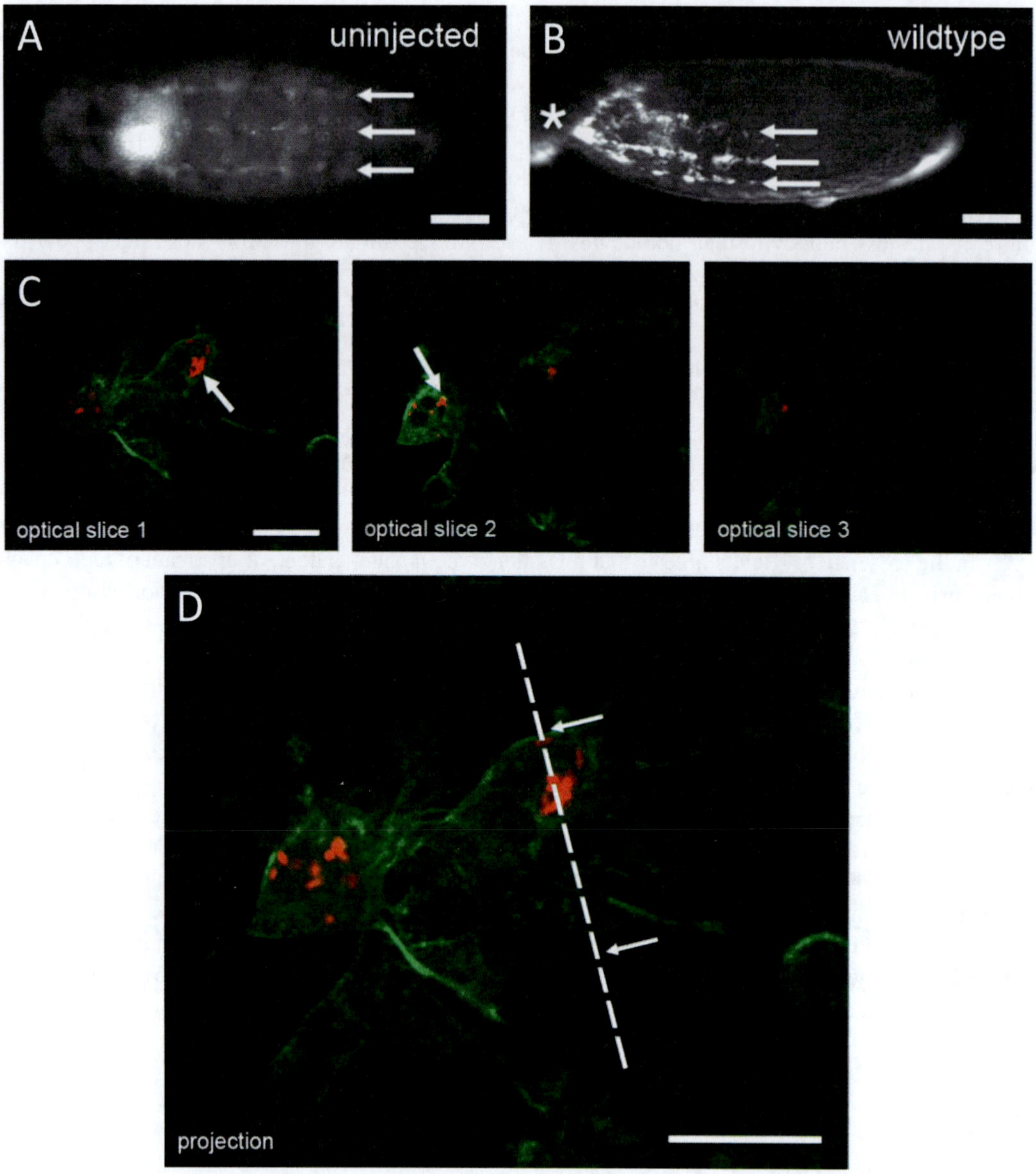

Fig. 6.1 Injected E. coli are recognised and engulfed by embryonic hemocytes (Vlisidou et al. 2009). (**a**) An embryo expressing GFP specifically in the hemocytes shows their characteristic pattern of distribution whereby the cells are arranged into three parallel lines running along the ventral surface of a stage 15 embryo (*arrows*). (**b**) Embryo injected with fluorescently labelled *E. coli* in the anterior region of the embryo (*asterisk*). Twenty minutes after injection hemocytes become labelled as they bind the injected bacteria (*arrows*). (**c**) Confocal images showing a series of optical slices taken through GFP expressing hemocytes. Images clearly show that the cells have internalised injected RFP labelled *E .coli* (*arrows*). (**d**) A projection of the slices shown in (**c**) highlight the two hemocytes (*green*) containing *E .coli* (*red*). *Arrows* mark the cell extremities. Scale bars represent 50 μm (**a** and **b**) and 10 μm (**c** and **d**)

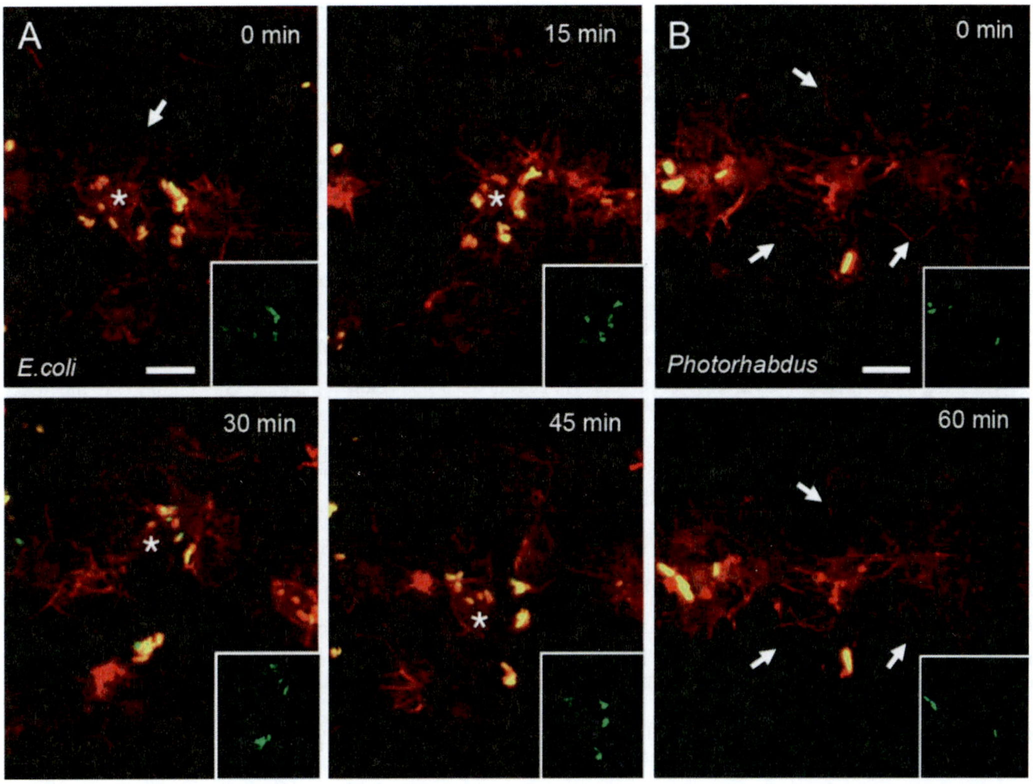

Fig. 6.2 Photorhabdus injection causes a dramatic freezing of embryonic hemocytes (Vlisidou et al. 2009). (**a**) Stills from a movie of hemocytes expressing RFPmoesin following injection of GFP expressing *E. coli*. Hemocytes (*asterisk*) can be clearly seen actively migrating within the embryo and extending dynamic actin rich protrusions (*arrow*) as they bind and clear the injected *E. coli* (*yellow*). Insets show the movement of bacteria over the period of the movie as they are carried within the migrating hemocytes. (**b**) Stills from a movie of RFP labelled hemocytes following injection of GFP labelled *Photorhabdus*. Hemocytes are able to recognise and bind the bacteria (*yellow*) but are completely frozen remaining in the same position for the duration of the movie (60 min). Cells still display actin rich protrusions (*arrows*) but these are static and bear no resemblance to the dynamic structures normally seen in hemocytes (compare to (**a**)). *Insets* show that the *Photorhabdus* bacteria themselves do not move throughout the movie as they are anchored to the static hemocytes. Scale bars represent 10 μm. Elapsed time is indicated in the *upper right* corner

References

Bataillé L, Augé B, Ferjoux G, Haenlin M, Waltzer L (2005) Resolving embryonic blood cell fate choice in Drosophila: interplay of GCM and RUNX factors. Development 132(20):4635–4644

Blow NS, Salomon RN, Garrity K, Reveillaud I, Kopin A et al (2005) Vibrio cholerae infection of Drosophila melanogaster mimics the human disease cholera. PLoS Pathog 1:e8

Brandt SM, Dionne MS, Khush RS, Pham LN, Vigdal TJ et al (2004) Secreted bacterial effectors and host-produced Eiger/TNF drive death in a Salmonella-infected fruit fly. PLoS Biol 2:e418

Brugirard-Ricaud K, Duchaud E, Givaudan A, Girard PA, Kunst F, Boemare N, Brehélin M, Zumbihl R (2005) Site-specific antiphagocytic function of the Photorhabdus luminescens type III secretion system during insect colonization. Cell Microbiol 7(3):363–371

Burridge K, Wennerberg K (2004) Rho and Rac take center stage. Cell 116(2):167–179

Ciche TA, Ensign JC (2003) For the insect pathogen Photorhabdus luminescens, which end of a nematode is out? Appl Environ Microbiol 69(4):1890–1897

Ciche TA, Kim KS, Kaufmann-Daszczuk B, Nguyen KC, Hall DH (2008) Cell invasion and matricide during Photorhabdus luminescens transmission by Heterorhabditis bacteriophora nematodes. Appl Environ Microbiol 74(8):2275–2287

Daborn PJ, Waterfield N, Silva CP, Au CP, Sharma S, Ffrench-Constant RH (2002) A single *Photorhabdus* gene, makes caterpillars floppy (mcf), allows *Escherichia coli* to persist within and kill insects. Proc Natl Acad Sci USA 99(16):10742–10747

Dong Y, Taylor HE, Dimopoulos G (2006) AgDscam, a hypervariable immunoglobulin domain-containing receptor of the Anopheles gambiae innate immune system. PLoS Biol 4(7):e229

Dowling AJ, Daborn PJ, Waterfield NR, Wang P, Streuli CH et al (2004) The insecticidal toxin makes caterpillars floppy (Mcf) promotes apoptosis in mammalian cells. Cell Microbiol 6:345–353

Dowling AJ, Waterfield NR, Hares MC, Le Goff G, Streuli CH et al (2007) The Mcf1 toxin induces apoptosis via the mitochondrial pathway and apoptosis is attenuated by mutation of the BH3-like domain. Cell Microbiol 9:2470–2484

Eleftherianos I, Millichap PJ, ffrench-Constant RH, Reynolds SE (2006a) RNAi suppression of recognition protein mediated immune responses in the tobacco hornworm Manduca sexta causes increased susceptibility to the insect pathogen Photorhabdus. Dev Comp Immunol 30(12):1099–1107

Eleftherianos I, Marokhazi J, Millichap PJ, Hodgkinson AJ, Sriboonlert A, ffrench-Constant RH, Reynolds SE (2006b) Prior infection of Manduca sexta with non-pathogenic Escherichia coli elicits immunity to pathogenic Photorhabdus luminescens: roles of immune-related proteins shown by RNA interference. Insect Biochem Mol Biol 36(6):517–525

Eleftherianos I, Boundy S, Joyce SA, Aslam S, Marshall JW, Cox RJ, Simpson TJ, Clarke DJ, ffrench-Constant RH, Reynolds SE (2007) An antibiotic produced by an insect-pathogenic bacterium suppresses host defenses through phenoloxidase inhibition. Proc Natl Acad Sci USA 104(7):2419–2424

Etienne-Manneville S, Hall A (2002) Rho GTPases in cell biology. Nature 420(6916):629–635

Farmer JJ 3rd, Jorgensen JH, Grimont PA, Akhurst RJ, Poinar GO Jr, Ageron E, Pierce GV, Smith JA, Carter GP, Wilson KL et al (1989) Xenorhabdus luminescens (DNA hybridization group 5) from human clinical specimens. J Clin Microbiol 27(7):1594–1600

Finlay BB, Falkow S (1989) Common themes in microbial pathogenicity. Microbiol Rev 53(2):210–230

Fischer-Le Saux M, Viallard V, Brunel B, Normand P, Boemare NE (1999) Polyphasic classification of the genus Photorhabdus and proposal of new taxa: P. Luminescens subsp. luminescens subsp. nov., P. luminescens subsp. akhurstii subsp. nov., P. luminescens subsp. laumondii subsp. nov., P. temperata sp. nov., P. Temperate subsp. temperata subsp. nov. and P. asymbiotica sp. nov. Int J Syst Bacteriol 49 Pt 4:1645–1656

Fleming V, Feil E, Sewell AK, Day N, Buckling A et al (2006) Agr interference between clinical Staphylococcus aureus strains in an insect model of virulence. J Bacteriol 188:7686–7688

Forst S, Dowds B, Boemare N, Stackebrandt E (1997) Xenorhabdus and Photorhabdus spp.: bugs that kill bugs. Annu Rev Microbiol 51:47–72

Frandsen JL, Gunn B, Muratoglu S, Fossett N, Newfeld SJ (2008) Salmonella pathogenesis reveals that BMP signaling regulates blood cell homeostasis and immune responses in Drosophila. Proc Natl Acad Sci USA 105:14952–14957

Gerrard JG, Vohra R, Nimmo GR (2003) Identification of Photorhabdus asymbiotica in cases of human infection. Commun Dis Intell 27(4):540–541

Gerrard J, Waterfield N, Vohra R, ffrench-Constant R (2004) Human infection with *Photorhabdus asymbiotica*: an emerging bacterial pathogen. Microbes Infect 6:229–237

Gillespie JP, Kanost MR, Trenczek T (1997) Biological mediators of insect immunity. Annu Rev Entomol 42:611–643

Hallem EA, Rengarajan M, Ciche TA, Sternberg PW (2007) Nematodes, bacteria, and flies: a tripartite model for nematode parasitism. Curr Biol 17(10):898–904, Epub 3 May 2007

Holz A, Bossinger B, Strasser T, Janning W, Klapper R (2003) The two origins of hemocytes in Drosophila. Development 130(20):4955–4962

Jacinto A, Wood W, Balayo T, Turmaine M, Martinez-Arias A, Martin P (2000) Dynamic actin-based epithelial adhesion and cell matching during Drosophila dorsal closure. Curr Biol 10(22):1420–1426

Jacinto A, Wood W, Woolner S, Hiley C, Turner L, Wilson C, Martinez-Arias A, Martin P (2002) Dynamic analysis of actin cable function during Drosophila dorsal closure. Curr Biol 12(14):1245–1250

Kim Y, Ji D, Cho S, Park Y (2005) Two groups of entomopathogenic bacteria, Photorhabdus and Xenorhabdus, share an inhibitory action against phospholipase A2 to induce host immunodepression. J Invertebr Pathol 89(3):258–264

Kim SH, Park SY, Heo YJ, Cho YH (2008) Drosophila melanogaster-based screening for multihost virulence factors of Pseudomonas aeruginosa PA14 and identification of a virulence-attenuating factor, HudA. Infect Immun 76:4152–4162

Kitamoto T (2001) Conditional modification of behaviour in Drosophila by targeted expression of a temperature-sensitive shibire allele in defined neurons. J Neurobiol 47:81–92

Mandal L, Dumstrei K, Hartenstein V (2004) Role of FGFR signaling in the morphogenesis of the *Drosophila* visceral musculature. Dev Dyn 231(2):342–348

Olofsson B, Page DT (2005) Condensation of the central nervous system in embryonic Drosophila is inhibited by blocking hemocyte migration or neural activity. Dev Biol 279(1):233–243

Paladi M, Tepass U (2004) Function of Rho GTPases in embryonic blood cell migration in Drosophila. J Cell Sci 117(Pt 26):6313–6326

Peel MM, Alfredson DA, Gerrard JG, Davis JM, Robson JM, McDougall RJ, Scullie BL, Akhurst RJ (1999) Isolation, identification, and molecular characterization of strains of Photorhabdus luminescens from infected humans in Australia. J Clin Microbiol 37(11):3647–3653

Pielage JF, Powell KR, Kalman D, Engel JN (2008) RNAi screen reveals an Abl kinase-dependent host cell pathway involved in Pseudomonas aeruginosa internalization. PLoS Pathog 4(3):e1000031

Shirasu-Hiza MM, Schneider DS (2007) Confronting physiology: how do infected flies die? Cell Microbiol 9(12):2775–2783

Silva CP, Waterfield NR, Daborn PJ, Dean P, Chilver T, Au CP, Sharma S, Potter U, Reynolds SE, ffrench-Constant RH (2002) Bacterial infection of a model insect: Photorhabdus luminescens and Manduca sexta. Cell Microbiol 4(6):329–339

Stramer B, Wood W, Galko MJ, Redd MJ, Jacinto A, Parkhurst SM, Martin P (2005) Live imaging of wound inflammation in Drosophila embryos reveals key roles for small GTPases during in vivo cell migration. J Cell Biol 168(4):567–573

Stuart LM, Boulais J, Charriere GM, Hennessy EJ, Brunet S, Jutras I, Goyette G, Rondeau C, Letarte S, Huang H, Ye P, Morales F, Kocks C, Bader JS, Desjardins M, Ezekowitz RA (2007) A systems biology analysis of the *Drosophila* phagosome. Nature 445(7123):95–101

Tepass U, Fessler LI, Aziz A, Hartenstein V (1994) Embryonic origin of hemocytes and their relationship to cell death in Drosophila. Development 120:1829–1837

Vlisidou I, Dowling AJ, Evans IR, Waterfield N, ffrench-Constant RH, Wood W (2009) Drosophila embryos as model systems for monitoring bacterial infection in real time. PLoS Pathog 5(7):e1000518

Waterfield NR, Sanchez-Contreras M, Eleftherianos I, Dowling A, Yang G, Wilkinson P, Parkhill J, Thomson N, Reynolds SE, Bode HB, Dorus S, ffrench-Constant RH (2008) Rapid Virulence Annotation (RVA): identification of virulence factors using a bacterial genome library and multiple invertebrate hosts. Proc Natl Acad Sci USA 105(41):15967–15972

Waterfield NR, Ciche T, Clarke D (2009) Photorhabdus and a host of hosts. Annu Rev Microbiol 63:557–574

Watson FL, Püttmann-Holgado R, Thomas F, Lamar DL, Hughes M, Kondo M, Rebel VI, Schmucker D (2005) Extensive diversity of Ig-superfamily proteins in the immune system of insects. Science 309(5742):1874–1878

Wojtowicz WM, Flanagan JJ, Millard SS, Zipursky SL, Clemens JC (2004) Alternative splicing of Drosophila Dscam generates axon guidance receptors that exhibit isoform-specific homophilic binding. Cell 118(5):619–633

Wojtowicz WM, Wu W, Andre I, Qian B, Baker D, Zipursky SL (2007) A vast repertoire of Dscam binding specificities arises from modular interactions of variable Ig domains. Cell 130(6):1134–1145

Wood W, Jacinto A (2007) Drosophila melanogaster embryonic haemocytes: masters of multitasking. Nat Rev Mol Cell Biol 8(7):542–551

Woolner S, Jacinto A, Martin P (2005) The small GTPase Rac plays multiple roles in epithelial sheet fusion–dynamic studies of Drosophila dorsal closure. Dev Biol 282(1):163–173

Chapter 7
Ecological Niche Modeling as a Tool for Understanding Distributions and Interactions of Vectors, Hosts, and Etiologic Agents of Chagas Disease

Jane Costa and A. Townsend Peterson

Abstract Chagas disease, or American Trypanosomiasis, is a tropical parasitic disease caused by the flagellate protozoan *Trypanosoma cruzi*, which is in turn transmitted by blood-sucking insects of the subfamily Triatominae (family Reduviidae). Because no drugs or vaccines are available to cure Chagas disease in its chronic phase, vectorial control (i.e., insecticide spraying) constitutes the principal means by which to impair Chagas disease transmission. Environmental and social factors have caused changes in the epidemiology of this disease—it was originally restricted to Latin America, but is now becoming a global heath concern in non-endemic areas as a consequence of human migrations. In Brazil, despite the fact that the most effective vector has been controlled, other triatomine species infest and colonize domiciliary habitats and can transmit the pathogen. As a consequence, Chagas disease transmission continues: the prevalence of the disease remains at ~12 million people, with ~200,000 new cases per year in 15 countries of Latin America, making control actions still necessary. Understanding the environmental requirements and geographic distributions of vectors is key to guiding control measures, and understanding better epidemiologic aspects of the disease. Ecologic niche modeling is a relatively new tool that permits such insights—as a consequence, here, we present an overview of insights gained using this approach in understanding of Chagas disease.

Introduction

The epidemiology of Chagas disease or American trypanosomiasis (Chagas 1909) has changed rapidly and significantly in recent decades because of environmental changes and human social factors (Briceno-Leon and Galvan 2007). This disease, originally restricted to the Latin America, is now becoming a global heath concern in non-endemic areas, owing to human migration to several developed countries (Schmunis and Yadon 2010). Chagas is a tropical parasitic disease caused by the flagellate protozoan *Trypanosoma cruzi*, which is transmitted by blood-sucking insects of the

J. Costa (✉)
Laboratório de Biodiversidade Entomológica, Instituto Oswaldo Cruz, Fiocruz, Rio de Janeiro, Brazil
e-mail: jcosta@ioc.fiocruz.br

A.T. Peterson
Biodiversity Institute, University of Kansas, Lawrence, KS 66045, USA
e-mail: town@ku.edu

E. Mylonakis et al. (eds.), *Recent Advances on Model Hosts*,
Advances in Experimental Medicine and Biology 710, DOI 10.1007/978-1-4419-5638-5_7,

subfamily Triatominae (family Reduviidae). Transmission of the pathogen to humans and other vertebrates occurs during or after the blood meal, through contaminated drops of feces of the vector species deposited on the skin of the host. Transmission may also occur via blood transfusion, organ transplantation, ingestion of food contaminated with parasites, and across the placenta. The most important vectors belong to the genera *Triatoma*, *Rhodnius*, and *Panstrongylus*. Currently, 141 triatomine species are known. Of this, 61 are present in Brazil (Lent and Wygodzinsky 1979; Galvão et al. 2003; Costa and Lorenzo 2009; Schofield and Galvão 2009), making Chagas disease a particular concern for Brazilian public health.

The South Cone Countries Initiative was launched in 1991, and included Brazil, Argentina, Paraguay, Uruguay, and Bolívia. The main objective of this international consortium was reduction of vectorial transmission of Chagas disease through insecticide control of *Triatoma infestans*, the most important vector, and by screening of blood donors. After nearly 10 years of the control action, incidence of the disease in those countries has been reduced by about 94% (WHO 2007). Subsequent initiatives were launched, including the Andean (1997), Central American Countries (1998), and Amazon (2004) efforts, which were aimed mainly at *Rhodnius prolixus*, the second most important vector (Guhl 2007). However, despite the successful metrics resulting from these control programs, the prevalence of the disease remains around 12 million people, with 200,000 new cases per year in 15 countries of Latin America (Morel and Lazdins 2003) (Fig. 7.1).

In Brazil, Chagas disease transmission still occurs – after the control measures against *T. infestans*, other species previously of secondary importance began infesting human habitations and serving as Chagas disease vectors. Importantly, most of this set of species are native to the Brazilian landscape (in contrast to *T. infestans*), and occur broadly under sylvatic conditions (also unlike *T. infestans*). As a consequence, while *T. infestans* could be eliminated efficiently from human environments, with a good degree of permanence, reinfestation of houses by native triatomine species is frequent and much more difficult to monitor and control (Silveira and Vinhaes 1999).

In Brazil, the vectors of greatest public health concern are those of the *T. brasiliensis* complex, in particular *T. b. brasiliensis* (Costa et al. 2003, 2009). Environmental changes seem to be

Fig. 7.1 *Triatoma infestans* the main Chagas disease vector (Picture Rodrigo Mexas/IOC/Fiocruz)

accompanied by rapid adaptive responses of these species. *T. sherlocki* is a good example of how fast vectors are able to change their behavior – the species was described in 2002 as an exclusively sylvatic vector, but recently it was found colonizing houses in remote sectors of Bahia state in an area newly colonized by humans (Almeida et al. 2009).

Brazil's Chagas disease control program, run by the Fundação Nacional de Saúde (FUNASA), has found that rapid ecological shifts can occur in response to control measures, demanding constant entomologic surveillance. For example, in Rio Grande do Sul, Brazil, after implementation of *T. infestans* control measures, *T. rubrovaria* became the most frequently found species inside human domiciles, although it was previously known only in sylvatic and peridomiciliary areas (Lent 1942). Indeed, *T. rubrovaria* now presents remarkable invasive behavior (Almeida et al. 2000). Hence, understanding potential ecological and geographic distributions of each potential vector species is key to directing effective control measures and understanding epidemiologic aspects of the disease better. Ecological niche modeling is a relatively new tool that permits researchers to explore geographic and ecologic phenomena based on known occurrences of species (Peterson 2006a). In this paper, we summarize some key contributions to understanding of Chagas disease etiology from this new set of approaches.

The Ecologic Niche Model (ENM) Concept

Joseph Grinnell (1917) originated the concept of ecologic niches, and was the first to explore connections between ecological niches and geographic distributions of species. His idea, translated into modern terminology, was that the ecologic niche of a species is the set of conditions under which it can maintain populations without immigration of individuals from other areas. A more complete discussion of the concept of ecologic niches and how they map ecologic requirements onto geographic distributions has been provided elsewhere (Soberón 2010) (Fig. 7.2).

The idea behind niche modeling is that known occurrences of species across landscapes, can be related to raster geographic information system coverages summarizing environmental variation across those landscapes, to estimate the ecological niche of the species (Peterson et al. 2002b). To the extent that it is successful in identifying the niche – or the habitable set of conditions for the species – ENM can be used to identify potential distributional areas for species on any landscape, which may include unsampled or unstudied portions of the native landscape (López-Cárdenas et al. 2005), areas of actual or potential invasion by species with expanding ranges (Peterson 2003), or changing potential distributional areas as a consequence of change (e.g., land use change or climate change; Sarkar et al. 2010) (Fig. 7.3).

ENM may be used to characterize distributional areas of species in complex, linked geographic and ecologic spaces. Specifically, ENMs permit researchers to characterize ecological needs of species, interpolate between sampling points to predict full distributions of species, predict species distribution into broadly unsampled areas, predict invasive potential in other regions/other continents, predict likely distributional change with changing land use, predict likely distributional change with changing climates, and build scenarios for understanding and characterizing unknown disease behavior. Hence, ENM offers a powerful tool for characterizing ecologic and geographic distributions of species across real-world landscapes (Peterson 2007, 2008a, b).

Numerous quantitative approaches to ENM have been explored (Elith et al. 2006), although frequently without appropriate consideration for what constitutes a "good" versus a "bad" model (Peterson et al. 2008). GARP and Maxent are the two methods that have seen broadest exploration and application to disease systems – both use evolutionary-computing approaches to detect and characterize associations between known occurrences and environmental characteristics of regions (Stockwell and Peters 1999; Phillips et al. 2006). In each case, the researcher supplies occurrence

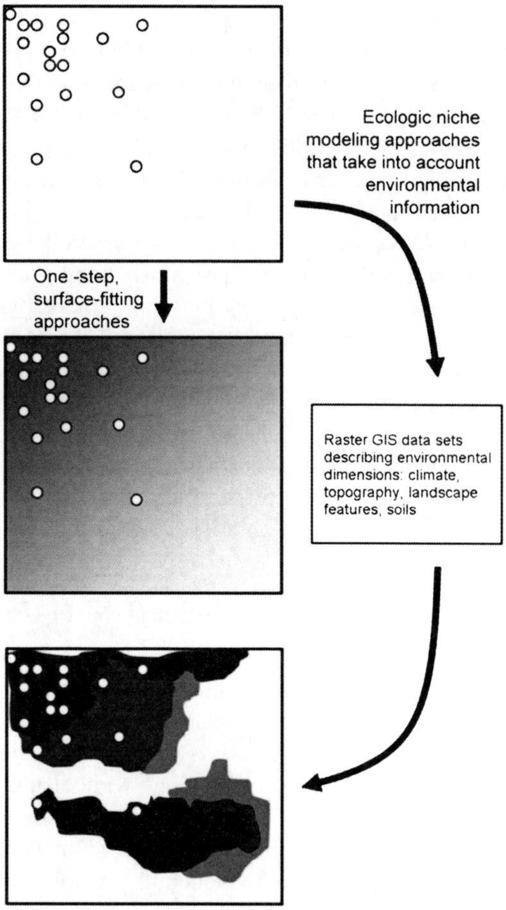

Fig. 7.2 Hypothetical example of a species' known occurrences (*circles*) and inferences from that information. The *middle panel* shows the pattern that would result from a surface-fitting or smoothing algorithm, and the *bottom panel* shows the ability of ecologic niche modeling approaches to detect unknown patterns in biologic phenomena based on the relationship between known occurrences and spatial patterns in environmental parameters. GIS, geographic information system (Reproduced from Peterson 2006a)

points and raster GIS data layers that summarize relevant environmental parameters, and both programs are available freely over the Internet. Model outputs include some degree of characterization of the form of the species' response to environmental conditions, and raster GIS summaries of areas falling within the ecological niche of the species (see much more complete summary of the details of the methodology in Peterson et al. 2011).

ENM and Chagas Disease

Ecologic niche modeling has been applied over the past decade to illuminate aspects of Chagas disease transmission. In this section, we highlight some of the more important contributions, and outline future steps that would improve an understanding of transmission still further.

Characterizing ecological niches – The first application of this methodology to infectious diseases was an assessment of niche differentiation among four members of the *T. brasiliensis* species complex (Costa et al. 2002). Considerable debate had existed regarding the nature of the distinct morphotypes of the *T. brasiliensis* complex. In particular, debates were raging as to whether the various *T. brasiliensis* populations represent morphs of a single, broadly distributed species, or whether they represent distinct biologic entities that could have distinct roles in ecological communities and in disease transmission cycles (Figs. 7.4 and 7.5).

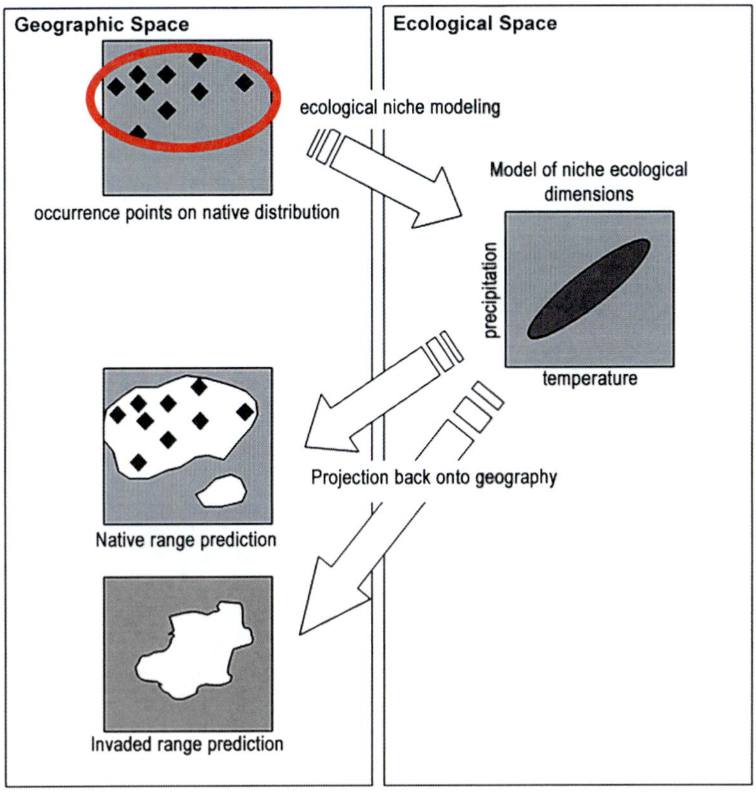

Fig. 7.3 Summary of main processes of ecologic niche modeling (Reproduced from Peterson 2006b)

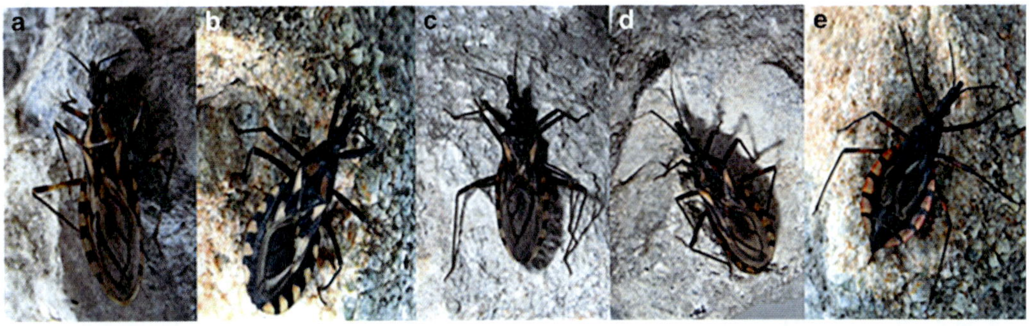

Fig. 7.4 Species and subspecies of the *Triatoma brasiliensis* species complex (Costa et al. 2009; Mendonça et al. 2009): (**a**) *Triatoma brasiliensis brasiliensis*; (**b**) *T. melanica*; (**c**) *T. juazeirensis*; (**d**) *T. b. macromelasoma*; (**e**) *T. sherlocki*

ENM offered a unique insight – that of rangewide differences among species in environmental dimensions. Four independent niche models were developed – one for each member of the complex. These models were then analyzed to test for ecologic differences among species and subspecies, as currently defined (Costa et al. 2006; Costa and Felix 2007). These models confirmed four ecologically distinct and differentiated entities within the complex, and allowed characterization of the dimensions in which their niches differed. What is more, patterns of ecologic similarity paralleled

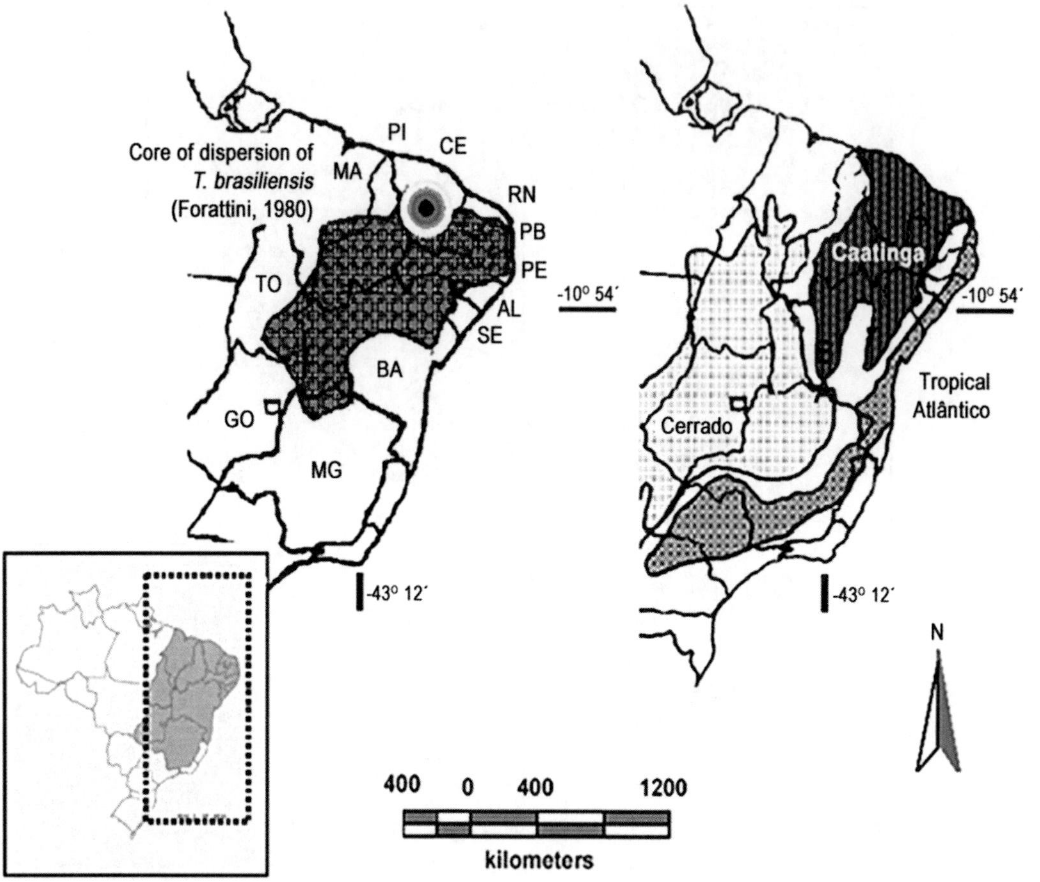

Fig. 7.5 The geographic area in which *Triatoma brasiliensis* species complex were captured in domiciliary ecotopes, according to Brazilian National Health Foundation (*AL* Alagoas, *BA* Bahia, *CE* Ceará, *GO* Goiás, *MA* Maranhão, *MG* Minas Gerais, *PB* Paraíba, *PE* Pernambuco, *PI* Piauí, *RN* Rio Grande do Norte, *SE* Sergipe, *TO* Tocantins) (Reproduced from Costa et al. 2003)

patterns of molecular differentiation (Costa et al. 1997b; Monteiro et al. 2004), suggesting that the distinct "populations" of the *T. brasiliensis* constituted a species complex including four forms with distinct epidemiological importance. These niche characterizations were also illuminating in defining niche characteristics for *T. b. macromelasoma*, of possible homoploidal hybrid origin. The ENM showed an ecologic space for *T. b. macromelasoma* not shared by either of the putative parental species, which bolstered the case for its hybrid origin (Costa et al. 1997a, b, 2009; Monteiro et al. 2004). A fifth member of this group (*T. sherlocki*) was revealed based on molecular tools (Mendonça et al. 2009), and ecological aspects of its distributional pattern were later characterized using ENM (Almeida et al. 2009) (Fig. 7.6).

Filling in unknown distributional information – The development of fine-scale predictions of potential distributional areas for each of the five members of the *T. brasiliensis* complex has allowed mapping distributional areas in considerable detail. When challenged with the task of anticipating the distribution of the species across broad, unsampled landscape, the models showed good predictive ability, well above null expectations, for all five members of the complex (Costa et al. 2010, in preparation). This work particularly focused attention on *T. b. brasiliensis*, as this

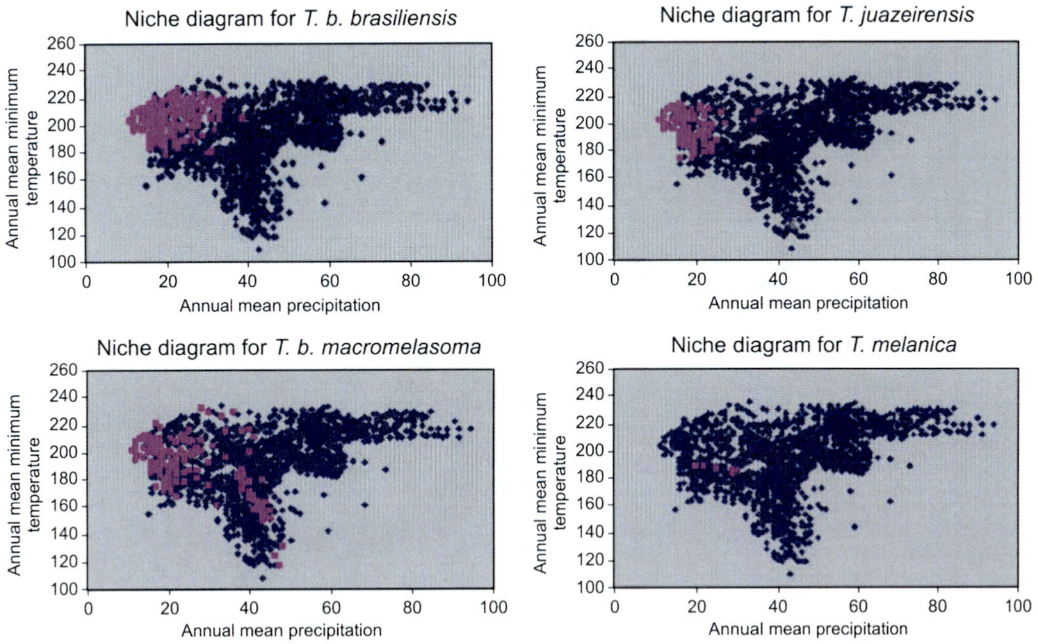

Fig. 7.6 Distribution of four populations of *T. brasiliensis* in ecologic space (annual mean minimum temperature vs annual mean precipitation, units multiplied by 10). Environmental combinations available in northeastern Brazil and adjacent areas are shown in *blue*, whereas those potentially inhabited by the members of the *T. brasiliensis* complex are shown in *pink* (Reproduced from Costa et al. 2002)

species appears to have the broadest environmental and geographic potential, which may open many opportunities to colonizing new areas (Costa et al. 2010, in preparation).

Other triatomine species, such as *Rhodnius nasutus* and *R. neglectus* in Brazil, have been investigated to estimate their potential distributions (Batista and Gurgel-Gonçalves 2009). Results supported the idea that the warmest and driest areas of northeastern Brazil offer ideal conditions for *R. nasutus* populations, whereas cold areas at high elevations with particular seasonal signatures of greenness indices (NDVI values) are more favorable for *R. neglectus* (Batista and Gurgel-Gonçalves 2009).

Sandoval-Ruiz et al. (2008) analyzed geographic distributions of three triatomine species in the Mexican state of Puebla. According to their models, the vector species *T. barberi* and *Meccus pallidipennis* are sympatric and widely distributed in the central-southern part of the state, whereas *T. dimidiata* is restricted to the northern mountains of the state, disjunct from the other species. In combination with information on human population distributions, *M. pallidipennis* was seen to present the highest vectorial competence values in the state. Finally, in broader analyses across North America, the *T. protracta* and *T. phyllosoma* complexes were found to occupy the broadest niches, which can translate into greater vectorial potential (Ibarra-Cerdeña et al. 2009).

Understanding host relationships – Elsewhere in Latin America, other challenges exist for understanding the full etiology of Chagas disease. In the southwestern United States and northern and central Mexico, the *T. protracta* complex is apparently associated with packrat (*Neotoma* spp.) species as hosts. At the time of the analysis, that species group was considered to include seven species (*T. protracta*, *T. peninsularis*, *T. sinaloensis*, *T. neotomae*, *T. barberi*, *T. nitida*, and *T. incrassata*). *T. protracta* further includes five forms that may or may not merit species status (*T. p. protracta*, *T. p. woodi*, *T. p. navajoensis*, *T. p. zacatecensis*, and *T. p. nahuatlae*; Ryckman

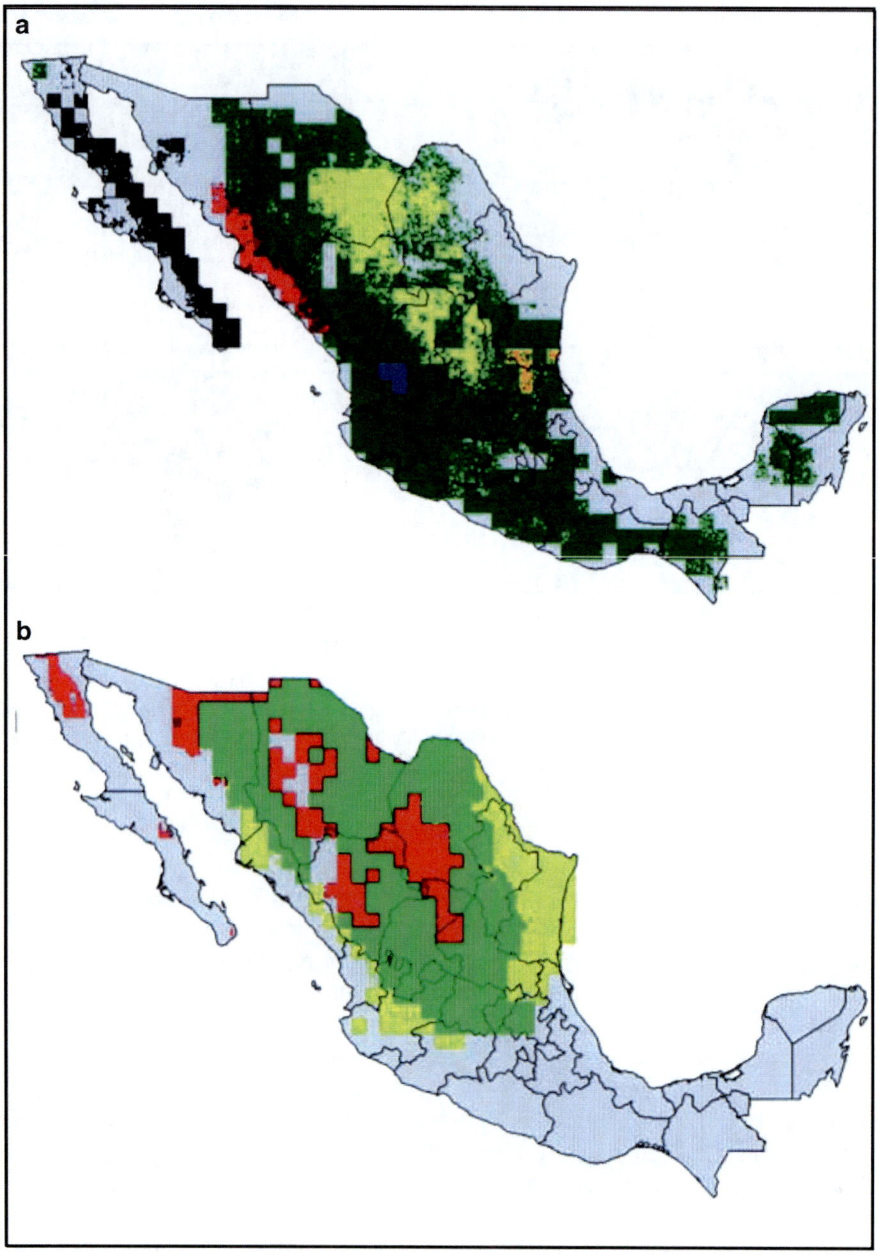

Fig. 7.7 Modeled geographic distributions of *Neotoma* woodrats in mainland Mexico. (**a**) *Black=Neotoma fuscipes, red=N. phenax, green=N. mexicana* (note that this distribution includes those of other, less widely distributed, species), *yellow=N. goldmani, blue=N. palatina, orange=N. angustapalata.* (**b**) *Red=N. albigula, yellow=N. micropus, green=N. albigula* and *N. micropus* (Reproduced from Peterson et al. 2002a)

1962). Early studies (Ryckman 1962) had posited high host specificity in this group, with each species associated with a species of *Neotoma* packrats. As host associations of *Triatoma* are often complex, this group was thus of considerable interest.

In an early ENM-based study exploring these ideas (Peterson et al. 2002a), niches and distributions were estimated for each species of *Triatoma* and *Neotoma*, and range overlap values calculated

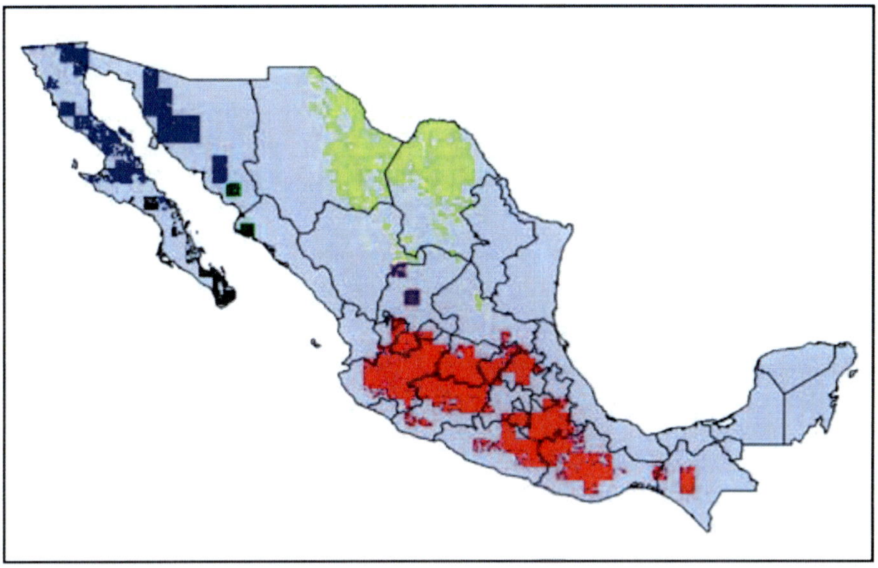

Fig. 7.8 Modeled geographic distributions of members of the *T. protracta* species group: *red*=*Triatoma barberi*, *yellow*=*T. p. woodi*, *green*=*T. sinaloensis*, *blue*=*T. p. protracta*, *black*=*T. peninsularis*, and *pink*=*T. p. zacatecensis*. Only areas predicted for each species at the highest level of confidence (all best-subsets models agree) are shown (Reproduced from Peterson et al. 2002a)

for each pairwise combination of bug and potential host. Species pairs showing highest overlap values in the ENM-based results coincided closely with those identified in much more intensive field studies (Ryckman 1962), suggesting that local interactions between *Triatoma* and *Neotoma* species and subspecies have implications at a geographic level, and that these interactions can potentially be reconstructed using ENM techniques (Peterson et al. 2002a) (Figs. 7.7 and 7.8).

A further example of illuminating transmission cycles of poorly known diseases comes from a study of Chagas disease in a domestic transmission cycle in southern Texas, USA. The study was initiated after three dogs died from acute Chagas cardiomyopathy. Studies were developed in both domiciliary and peridomiciliary ecotopes, including serologic studies on stray dogs. ENMs were developed to predict areas where the vector *T. gerstaeckeri* might be expected to occur. The result was a picture in which an insect species typically considered as a zoonotic vector had invaded domestic environments, establishing a transmission cycle centered on dogs. Further analyses are required to discover whether these findings represent an isolated case or an emerging public health problem, but the indication is of a potentially broad distribution for *T. gerstaeckeri* presenting potential public health risks (Beard et al. 2003) that will perhaps worsen with warming climates (Sarkar et al. 2010).

Pros, Cons and Conclusions

ENM has seen increasing use and application over the last 10 years thanks in large part to the broad applicability of this tool to understanding ecologic and geographic distributions of diverse sectors of biodiversity. Other studies have focused on aspects of malaria (Rosa-Freitas et al. 2007), leishmaniosis (Peterson and Shaw 2003), dengue (Peterson et al. 2005), blastomycosis (Reed et al. 2008),

Ebola virus (Lash et al. 2008) Marburg virus (Peterson et al. 2006), and other zoonotic diseases.. This methodology has seen extensive testing and sensitivity analysis (Peterson et al. 2011).

ENM is of course, not without its limitations. Indeed, as it has taken a more apparent role in the literature, and as tools for developing ENMs have become more easily available and more convenient to use, many applications have been developed without a firm conceptual basis. That is, any ecologic niche model should be developed in a context that is appropriate to such estimations: the subtle distinctions between ecological niche modeling and species distribution modeling must be borne in mind at all times (Peterson 2006b), such that models are appropriate to the challenges of estimation that are being confronted.

More practically, ENMs must always be developed carefully so as not to create discords and mismatches. That is, occurrence data and environmental data sets must coincide, both in terms of time span and in terms of spatial resolution and precision. If ecologic niches are the target of modeling, then the full geographic range must generally be assessed – in this way, the models have the best probability of characterizing the full dimensions of the niche. In sum, although ENM is now quite easy to perform computationally, the details of the process remain difficult, and a suite of complications exists that is frequently not appreciated.

Finally, looking forward, several dimensions of ENM application to disease-related challenges remain to be fully explored. Perhaps most intriguing is the comparison and contrasting of models based on human case occurrences, versus meta-models based on the integration of multiple individual models for reservoir and vector species. The latter approach reconstructs the entire transmission system from its components; the former approach should in theory arrive at the same end-product, but is based on the "integration" that is achieved by the transmission system that leads to humans. Parallel development of these two points of view for one disease system would provide very interesting insights.

Acknowledgements Vanessa Lima Neiva for editing the references and Heloisa Diniz, Serviço de Produção e Tratamento de Imagens do Instituto Owaldo Cruz, for editing the figures and assistance in the preparation of the plates. CNPq for support.

References

Almeida CE, Vinhaes MC, Almeida JR, Silveira AC, Costa J (2000) Monitoring the domiciliary and peridomiciliary invasion process of *Triatoma rubrovaria* in the state of Rio Grande do Sul, Brazil. Mem Inst Oswaldo Cruz 95:761–768

Almeida CE, Folly-Ramos E, Peterson AT, Lima-Neiva V, Gumiel M, Duarte R, Lima MM, Locks M, Beltrão M, Costa J (2009) Could *Triatoma sherlocki* be vectoring Chagas disease in small mining communities in Bahia, Brazil? Med Vet Entomol 23:410–417

Batista TA, Gurgel-Gonçalves R (2009) Ecological niche modelling and differentiation between *Rhodnius neglectus* Lent, 1954 and *Rhodnius nasutus* Stål, 1859 (Hemiptera: Reduviidae: Triatominae) in Brazil. Mem Inst Oswaldo Cruz 104:1165–1170

Beard CB, Pye G, Steurer FJ, Salinas Y, Campman R, Peterson AT, Ramsey JM, Wirtz RA, Robinson LE (2003) Chagas disease in a domestic transmission cycle in southern Texas, USA. Emerg Infect Dis 9:103–105

Briceno-Leon R, Galvan MJ (2007) The social determinants of Chagas disease and the transformations of Latin America. Mem Inst Oswaldo Cruz 102:109–112

Chagas C (1909) Nova tripanozomiaze humana: estudos sobre a morfologia e o ciclo evolutivo do *Schizotrypanum cruzi* n. gen. n. sp., agente etiologico de nova entidade morbida do homem. Mem Inst Oswaldo Cruz 1:159–218

Costa J, Felix M (2007) *Triatoma juazeirensis* sp. nov. from the state of Bahia, northeastern Brazil (Hemiptera: Reduviidae: Triatominae). Mem Inst Oswaldo Cruz 102:87–90

Costa J, Lorenzo M (2009) Biology, diversity and strategies for the monitoring and control of triatomines – Chagas disease vectors. Mem Inst Oswaldo Cruz 104(suppl):46–51

Costa J, Barth OM, Marchon-Silva V, Almeida CE, Freitas-Sibajev MG, Panzera F (1997a) Morphological studies on the *Triatoma brasiliensis* Neiva, 1911 (Hemiptera, Reduviidae, Triatominae) – genital structures and eggs of different chromatic forms. Mem Inst Oswaldo Cruz 92:493–498

Costa J, Freitas-Sibajev MG, Marchon-Silva V, Pires MQ, Pacheco R (1997b) Isoenzymes detect variation in populations of *Triatoma brasiliensis* (Hemiptera–Reduviidae–Triatominae). Mem Inst Oswaldo Cruz 92: 459–464

Costa J, Peterson AT, Beard CB (2002) Ecologic niche modeling and differentiation of populations of *Triatoma brasiliensis* Neiva, 1911, the most important Chagas disease vector in northeastern Brazil (Hemiptera, Reduviidae, Triatominae). Am J Trop Med Hyg 67:516–520

Costa J, Almeida CE, Dotson EM, Lins A, Vinhaes M, Silveira AC, Beard CB (2003) The epidemiologic importance of *Triatoma brasiliensis* as a Chagas disease vector in Brazil: a revision of domiciliary captures during 1993–1999. Mem Inst Oswaldo Cruz 98:443–449

Costa J, Argolo AM, Felix M (2006) Redescription of *Triatoma melanica* Neiva & Lent, 1941, new status (Hemiptera: Reduviidae: Triatominae). Zootaxa 1385:47–52

Costa J, Peterson AT, Dujardin JP (2009) Morphological evidence suggests homoploid hybridization as a possible mode of speciation in the Triatominae (Hemiptera: Heteroptera: Reduviidae). Infect Genet Evol 9:263–270

Costa J, Dornak L, Almeida CE, Peterson TA (2010) Fine scale predictions on *T. brasiliensis* complex. Angean conferences series, 52:55. I International conference on model host, Crete, Greece

Elith J, Graham CH, Anderson RP, Dudik M, Ferrier S, Guisan A, Hijmans RJ, Huettman F, Leathwick JR, Lehmann A, Li J, Lohmann LG, Loiselle BA, Manion G, Moritz C, Nakamura M, Nakazawa Y, Overton J, Peterson AT, Phillips SJ, Richardson KS, Scachetti-Pereira R, Schapire RE, Soberón J, Williams S, Wisz MS, Zimmermann NE (2006) Novel methods improve prediction of species' distributions from occurrence data. Ecography 29:129–151

Galvão C, Carcavallo R, Rocha DS, Jurberg J (2003) A checklist of the current valid species of the subfamily Triatominae Jeannel, 1919 (Hemiptera: Reduviidae) and their geografical distribution with nomenclatural and taxonomic notes. Zootaxa 202:1–36

Grinnell J (1917) Field tests of theories concerning distributional control. Am Naturalist 51:115–128

Guhl F (2007) Chagas disease in Andean countries. Mem do Inst Oswaldo Cruz 102(suppl 1):1–9

Ibarra-Cerdeña CN, Sánchez-Cordero V, Peterson AT, Ramsey JM (2009) Ecology of North American Triatominae. Acta Trop 110:178–186

Lash RL, Brunsell N, Peterson AT (2008) Spatiotemporal environmental triggers of Ebola and Marburg virus transmission. GeoCarto Int 23:451–466

Lent H (1942) Estudos sobre os triatomíneos do estado do Rio Grande do Sul, com descrição de uma espécie nova. Rev Bras Biol 2:219–231

Lent H, Wygodzinsky P (1979) Revision of the Triatominae (Hemiptera: Reduviidae) and their significance as vector of Chagas disease. Bull Am Mus Nat Hist 163:123–520

López-Cárdenas J, Gonzalez-Bravo FE, Salazar-Schettino PM, Gallaga-Solorzano JC, Ramírez-Barba E, Martínez-Méndez J, Sänchez-Cordero V, Peterson AT, Ramsey JM (2005) Fine-scale predictions of distributions of Chagas disease vectors in the state of Guanajuato, Mexico. J Med Entomol 42:1068–1081

Mendonça VJ, Silva MTA, Araújo RF, Martins Júnior J, Bacci Júnior M, Almeida CE, Costa J, Graminha MAS, Cicarelli RMB, Rosa JA (2009) Phylogeny of *Triatoma sherlocki* (Hemiptera: Reduviidae: Triatominae) inferred from two mitochondrial genes suggests its location within the *Triatoma brasiliensis* complex. Am J Trop Med Hyg 81:858–864

Monteiro FA, Donnelly MJ, Beard CB, Costa J (2004) Nested clade and phylogeographic analyses of the Chagas disease vector *Triatoma brasiliensis* in northeast Brazil. Mol Phylogenet Evol 32:46–56

Morel CM, Lazdins J (2003) Chagas disease. Nat Rev Microbiol 1:14–15

Peterson AT (2003) Predicting the geography of species' invasions via ecological niche modeling. Q Rev Biol 78:419–433

Peterson AT (2006a) Ecological niche modeling and spatial patterns of disease transmission. Emerg Infect Dis 12:1822–1826

Peterson AT (2006b) Uses and requirements of ecological niche models and related distributional models. Biodiv Inf 3:59–72

Peterson AT (2007) Ecological niche modelling and understanding the geography of disease transmission. Vet Ital 43:393–400

Peterson AT (2008a) Biogeography of diseases: a framework for analysis. Naturwissenschaften 45:483–491

Peterson AT (2008b) Improving methods for reporting spatial epidemiologic data. Emerg Infect Dis 14:1335–1337

Peterson AT, Shaw JJ (2003) *Lutzomyia* vectors for cutaneous leishmaniasis in southern Brazil: ecological niche models, predicted geographic distributions, and climate change effects. Int J Parasitol 33:919–931

Peterson AT, Sánchez-Cordero V, Beard CB, Ramsey JM (2002a) Ecological niche modeling and potential reservoirs for Chagas disease, Mexico. Emerg Infect Dis 8:662–667

Peterson AT, Stockwell DRB, Kluza DA (2002b) Distributional prediction based on ecological niche modeling of primary occurrence data. In: Scott JM (ed) Predicting species occurrences: issues of scale and accuracy. Island Press, Washington, D.C

Peterson AT, Martínez-Campos C, Nakazawa Y, Martínez-Meyer E (2005) Time-specific ecological niche modeling predicts spatial dynamics of vector insects and human dengue cases. Trans R Soc Trop Med Hyg 99:647–655

Peterson AT, Lash RR, Carroll DS, Johnson KM (2006) Geographic potential for outbreaks of Marburg hemorrhagic fever. Am J Trop Med Hyg 75:9–15

Peterson AT, Papes M, Soberón J (2008) Rethinking receiver operating characteristic analysis applications in ecological niche modelling. Ecol Model 213:63–72

Peterson AT, Soberón J, Pearson RG, Anderson RP, Martínez-Meyer E, Nakamura M, Araújo MB (2011) Ecological niches and geographic distributions. Princeton University Press, Princeton

Phillips SJ, Anderson RP, Schapire RE (2006) Maximum entropy modeling of species geographic distributions. Ecol Model 190:231–259

Reed KD, Meece JK, Archer JR, Peterson AT (2008) Ecologic niche modeling of *Blastomyces dermatitidis* in Wisconsin. PLoS One 3:e2034

Rosa-Freitas MG, Tsouris P, Peterson AT, Honório NA, Barros FSMD, Aguiar DBD, Gurgel HDC, Arruda MED, Vasconcelos SD, Luitgards-Moura JF (2007) An ecoregional classification for the state of Roraima, Brazil: the importance of landscape in malaria biology. Mem Inst Oswaldo Cruz 102:349–358

Ryckman RE (1962) Biosystematics and hosts of the *Triatoma* complex in North America. Univ Calif Publ Ent 27:93–239

Sandoval-Ruiz CA, Zumaquero-Rios JL, Rojas-Soto OR (2008) Predicting geographic and ecological distributions of triatomine species in the southern Mexican state of Puebla using ecological niche modeling. J Med Entomol 45:540–546

Sarkar S, Strutz SE, Frank DM, Rivaldi CL, Sissel B, Sánchez-Cordero V (2010) Chagas disease risk in Texas. PLoS Negl Trop Dis 4:e836

Schmunis GA, Yadon ZE (2010) Chagas disease: a Latin American health problem becoming a world health problem. Acta Trop 115:14–21

Schofield CJ, Galvão C (2009) Classification, evolution and species groups within the Triatominae. Acta Trop 110:88–100

Silveira AC, Vinhaes MC (1999) Elimination of vector-borne transmission of Chagas disease. Mem Inst Oswaldo Cruz 94(suppl I):405–411

Soberón J (2010) Niche and area of distribution modeling: a population ecology perspective. Ecography 33:159–167

Soberón J, Peterson AT (2007) Interpretation of models of fundamental ecological niches and species' distributional areas. Biodiv Inf 2:1–10

Stockwell DRB, Peters DP (1999) The GARP modelling system: problems and solutions to automated spatial prediction. Int J Geogr Inf Syst 13:143–158

World Health Organization (WHO) (2007) Disponível em: [http://www.who.org] Accessed in June de 2010

Chapter 8
Where Simplicity Meets Complexity: *Hydra*, a Model for Host–Microbe Interactions

René Augustin, Sebastian Fraune, Sören Franzenburg, and Thomas C.G. Bosch

Abstract For a long time, the main purpose of microbiology and immunology was to study pathogenic bacteria and infectious disease; the potential benefit of commensal bacteria remained unrecognised. Discovering that individuals from *Hydra* to man are not solitary, homogenous entities but consist of complex communities of many species that likely evolved during a billion years of coexistence (Fraune and Bosch 2010) led to the hologenome theory of evolution (Zilber-Rosenberg and Rosenberg 2008) which considers the holobiont with its hologenome as the unit of selection in evolution. Defining the individual microbe–host conversations in these consortia is a challenging but necessary step on the path to understanding the function of the associations as a whole. Untangling the complex interactions requires simple animal models with only a few specific bacterial species. Such models can function as living test tubes and may be key to dissecting the fundamental principles that underlie all host–microbe interactions. Here we introduce *Hydra* (Bosch et al. 2009) as such a model with one of the simplest epithelia in the animal kingdom (only two cell layers), with few cell types derived from only three distinct stem cell lineages, and with the availability of a fully sequenced genome and numerous genomic tools including transgenesis. Recognizing the entire system with its inputs, outputs and the interconnections (Fraune and Bosch 2010; Bosch et al. 2009; Fraune and Bosch 2007; Fraune et al. 2009a) we here present observations which may have profound impact on understanding a strictly microbe-dependent life style and its evolutionary consequences.

The Basal Metazoan Model Organism *Hydra* Enters the Genomic Era

Hydra belongs to one of the most basal eumetazoan phylum, the Cnidaria which are a sister taxon to all Bilateria. *Hydra* represents a classical model organism in developmental biology which was introduced by Abraham Trembley as early as 1744 (Trembley 1744). Because of its simple body plan, having only two epithelial layers (an endodermal and ectodermal epithelium separated by an extracellular matrix termed mesogloea), a single body axis with a head, gastric region and foot, and a limited number of different cell types, *Hydra* served for many years as model in developmental biology to approach basic mechanisms underlying de novo pattern formation, regeneration, and cell differentiation.

R. Augustin (✉) • S. Fraune • S. Franzenburg • T.C.G. Bosch
Zoological Institute, Christian-Albrechts-University Kiel, Olshausenstr. 40, Kiel 24098, Germany
e-mail: raugustin@zoologie.uni-kiel.de

E. Mylonakis et al. (eds.), *Recent Advances on Model Hosts*,
Advances in Experimental Medicine and Biology 710, DOI 10.1007/978-1-4419-5638-5_8,

The genome of *Hydra magnipapillata* is relatively large (1,300 Mb) (Chapman et al. 2010). Since up to 40% of the whole genome is composed of transposable elements (Chapman et al. 2010), this was interpreted as "a very dynamic genome" in which recombination events might occur even without sexual recombination. Whether this, in combination with horizontal gene transfer and trans-splicing, allows the immortal, constantly regenerating and asexually proliferating polyps to quickly adapt to changing environmental conditions, remains a matter of debate. In addition to the *Hydra magnipapillata* genome, a large set of expressed sequence tags (ESTs) is available at www.compagen.org (Hemmrich and Bosch 2008). Adding to the relatively rich data sets available in *Hydra*, in recent years, additional genome and transcriptome sequences have become available from related basal metazoans such as corals and *Nematostella,* shedding new and bright light on the ancestral gene repertoire (Putnam et al. 2007; Rast et al. 2006; Srivastava et al. 2008). The accumulated data show that Cnidaria posses most of the gene families found in bilaterians (Putnam et al. 2007; Kusserow et al. 2005; Kortschak et al. 2003; Miller et al. 2005) and therefore have retained many ancestral genes that have been lost in *D. melanogaster* and *C. elegans* (Miller et al. 2005; Technau et al. 2005). Since the genome organization and genome content of Cnidaria is remarkably similar to that of morphologically much more complex bilaterians, these animals offer unique insights into the content of the "genetic tool kit" present in the Cnidarian–bilaterian ancestor.

For analytical purposes, an important technical breakthrough in studies using basal metazoans was the development of a transgenic procedure allowing efficient generation of transgenic *Hydra* lines by embryo microinjection (Wittlieb et al. 2006). This not only allows functional analysis of genes controlling development and immune reactions, but also in vivo tracing of cell behavior.

Hydra Has an Effective Innate Immune System

The tube-like body structure of *Hydra* resembles in several aspects the anatomy of the vertebrate intestine with the endodermal epithelium lining the gastric cavity, and the ectodermal epithelium providing a permanent protection barrier to the environment (Fig. 8.1a, b). In *Hydra*, epithelial cells in both layers are multifunctional having both secretory and phagocytic activity (Bosch et al. 2009; Bosch and David 1986). A combined biochemical and transcriptome analysis approach revealed that in *Hydra,* most innate immune responses are mediated by epithelial cells (Bosch et al. 2009; Jung et al. 2009). Although there are no motile immune effector cells or phagocytes present in *sensu stricto*, endodermal epithelial cells not only contribute to digestion and uptake of food, but also are able to phagocytose bacteria present in the gastric cavity (Fig. 8.1c). Within the endodermal layer, gland cells contribute to innate immune reactions by producing potent antimicrobial serine protease inhibitors (Augustin et al. 2009a). Some cnidarians have a remarkable capability of regeneration. In *Hydra*, for example, gross damage to the tissue is quickly repaired due to the presence of continuously proliferating stem cells (Bosch 2007). Since cells infected or damaged by pathogens, such as bacteria or fungi, are quickly removed by apoptosis (Bosch and David 1986) and replaced by non-infected cells, this enormous regeneration capacity may be considered an additional arm of the innate immune defense.

At the molecular level *Hydra* recognizes "Microbial Associated Molecular Patterns" (MAMPs) with the help of the TLR signaling pathway. The Toll-like receptor in *Hydra*, functions as a co-receptor with the MAMP recognizing Leucin-rich repeats and the signal transmitting TIR domain on two separated but interacting proteins (Bosch et al. 2009) (Fig. 8.2). Homologous sequences to nearly all other components of the TLR pathway were identified (Bosch et al. 2009) including NFkB (unpublished) (Fig. 8.2). RNAi knock down experiments with Hydra TLR showed a drastic reduction of antimicrobial activity in the knock down tissue compared to the wild type, which makes it apparent that antimicrobial activity relies directly on the activation of the TLR cascade

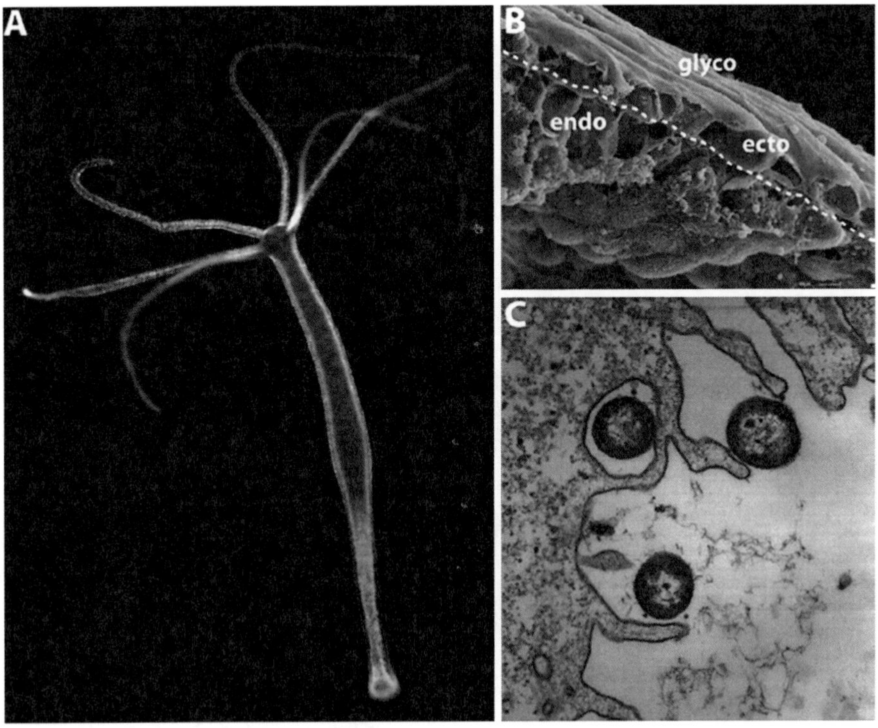

Fig. 8.1 Cnidarians are diploblastic animals. (**a**) Live image of *Hydra oligactis* (Photo by S. Fraune). (**b**) Raster electron micrograph showing the ectodermal (*ecto*) and endodermal epithelium (*endo*); which are separated by an extracellular matrix (mesoglea – *dashed line*); a true mesoderm is missing. The apical part of *Hydra* ectodermal epithelial cells is covered by a glycocalix layer (glyco) (Photo by F. Anton-Erxleben). (**c**) Transmission electron micrographic of endodermal epithelial cells phagocyte bacteria from the gastric lumen (Panel C modified from Bosch et al. 2009)

(Bosch et al. 2009) (Fig. 8.2). In addition to recognizing MAMPs at the cell membrane, intracellular recognition of bacteria in *Hydra* is mediated by an unexpected large number of cytosolic NOD-like receptors (Lange et al. 2011). We have proposed elsewhere (Lange et al. 2011) that upon their activation and dimerisation apoptosis as an evolutionary old mechanism in immune defence might be initiated (Lange et al. 2011).

Prominent effector molecules downstream of the conserved TLR cascade are antimicrobial peptides (AMPs). Up to now we have isolated four families of antimicrobial peptides including the Hydramacins, Arminins, Periculins and serine protease inhibitors of the Kazal type (Bosch et al. 2009; Jung et al. 2009; Augustin et al. 2009a, b; Fraune et al. 2010). Interestingly, all antimicrobial peptides isolated so far are present in endodermal tissue only. That supports the view that the endodermal epithelium surrounding the gastric cavity is especially endangered by the regular uptake of food, and that hydra's AMPs contribute to the chemical defense properties of this layer – similar to AMPs in the human small intestine. Periculins and Arminins are made as precursors. To activate them, a negatively charged N-terminal domain is cleaved and the highly positively charged C-terminal domain is released (Bosch et al. 2009; Augustin et al. 2009b). In the Periculin family, this cationic C-terminal region is rich in cysteines, indicating that this domain requires a distinct three dimensional structure for activity. In addition to endodermal epithelial cells, Periculin is also expressed in female germ cells and used for maternal protection of the embryo (Fraune et al. 2010).

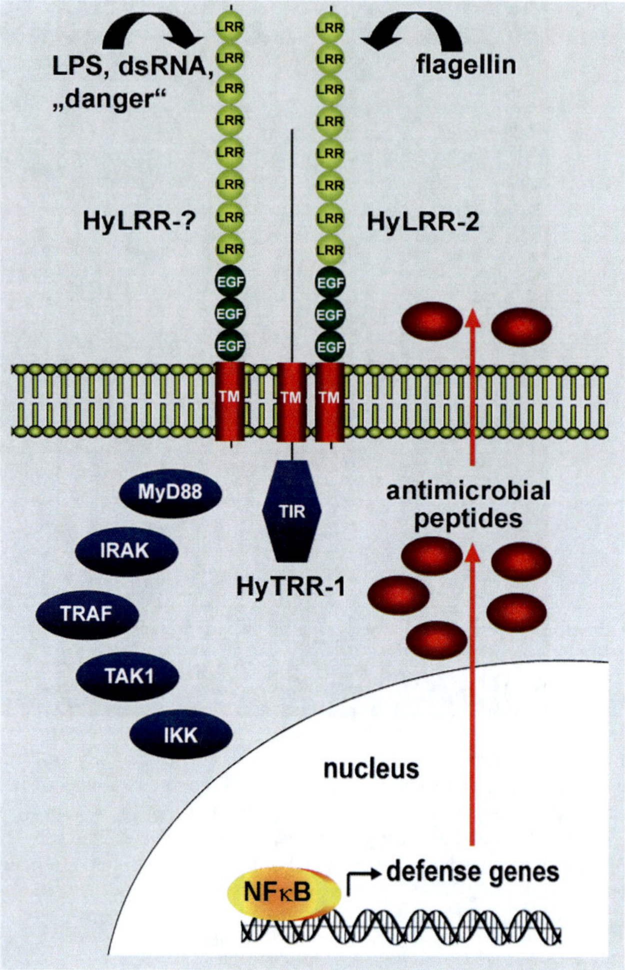

Fig. 8.2 Molecular components of the pathways involved in the *Hydra* epithelial host defence system (Panel modified from Bosch et al. 2009)

Arminins are characterized by a positively charged and variable C-terminus that includes the last 31 amino acids (pI of 12.1) (Augustin et al. 2009b). When used in liquid growth inhibition assays, the C-terminal part of Arminin 1a is capable of killing a large number of antibiotic resistant bacteria including methicillin resistant *S. aureus* (Fig. 8.3) and vancomycin resistant strains of *E. faecalis* and *E. faecium* (Augustin et al. 2009b).

Hydramacins have no precursor form. They are made as single peptides with an N-terminal signal peptide which is followed by an eight cysteine containing cationic C-terminal part. The peptide seems to be highly active mostly against gram negative bacteria also including human pathogenic multi resistant bacteria (Bosch et al. 2009). The three dimensional structure is characterized by an unusual arrangement of cationic and hydrophobic amino acid residues. The cationic amino acids form a central ring that is flanked by two hydrophobic parts. Based on a 3D model we hypothesize that bacteria attach to hydramacin to form large aggregates (Jung et al. 2009). The model is supported by an observation that after application of Hydramacin-1 bacteria aggregate, precipitate and finally die. Electron microscopic pictures show that these bacteria die within intact membranes (Jung et al. 2009). We assume that the aggregation processes may induce some kind of programmed cell death in bacteria (Engelberg-Kulka et al. 2005, 2006). If true, it might also explain

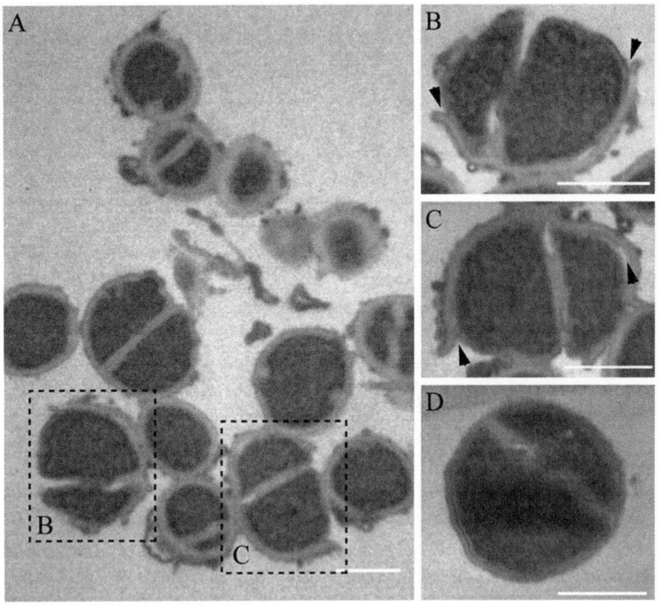

Fig. 8.3 Morphology changes of cArminin 1a treated *S. aureus* ATCC 12600. (**a**) Transmission electron micrograph of *S. aureus* (10^8 cells/mL) incubated with cArminin 1a for 1.5 h. (**b**, **c**) Magnification of two bacterial cells of (**a**); *arrows* point to the detachment of peripheral cell wall. (**d**) Transmission electron micrograph of *S. aureus* (10^8 cells/mL) incubated in 10 mM sodium phosphate buffer pH 7.4 for 1.5 h as negative control (intact cells). *Bars* represent 1 μm (Panel from Augustin et al. 2009b)

the very low doses at which Hydramacin-1 normally is active. Since human pathogens share little or no evolutionary history with Hydra-associated microbes and appear to be particularly vulnerable by hydra-antimicrobial molecules, antimicrobial peptides from such basal metazoans may provide interesting lead structures for a novel generation of antibiotics.

Taken together, *Hydra* has an effective innate immune system to interact with bacteria at the epithelial interface. The crucial question now is whether its main function is to keep out pathogens, or to allow the right community of microbes in.

Hydra and the Hologenome Theory of Evolution

In the traditional view of evolutionary biologists, the concept of individual selection posits that adaptation takes place on the level of individuals or genes. Wilson and Sober (1989) expanded this concept to the 'superorganism,' which considers selection on individuals (or genes), but additionally also on single- or multispecies communities. Based on the concept of the 'holobiont', Rosenberg and colleagues in 2007 proposed that organisms such as corals are able to adapt rapidly to changing environmental conditions by altering their associated microbiota. Depending on the variety of different niches provided by the host (which can change with the developmental stage, the diet or other environmental factors), a more or less diverse microbial community can establish with a given host species (Zilber-Rosenberg and Rosenberg 2008). Since this may provide corals, for example, with resistance against certain pathogens (Rosenberg et al. 2007), enabling them to adapt much faster to novel environmental conditions than by mutation and selection, host–microbe interactions must also be considered as significant drivers of animal evolution and diversification.

To test theories regarding the assembly of tissue-associated microbial communities and to gain insights into the function of the microbiota of a phylogenetically ancient epithelium, we have started

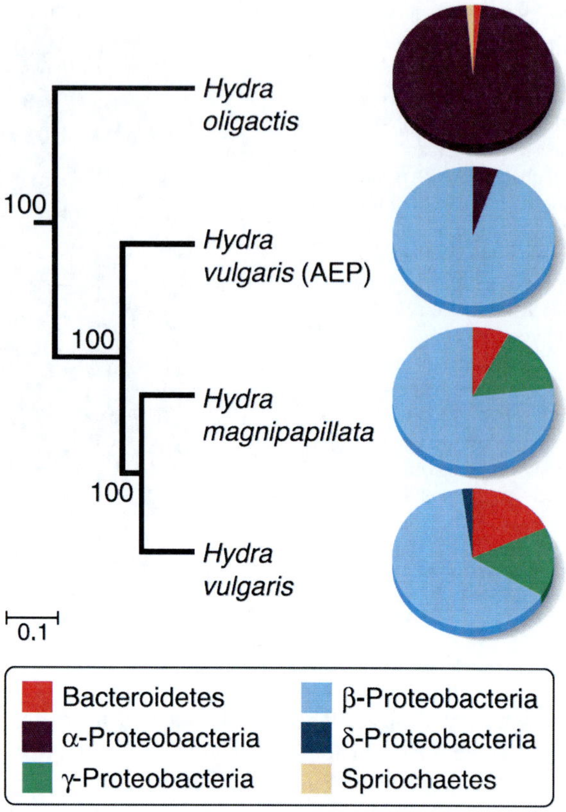

Fig. 8.4 *Hydra* polyps are colonized by species specific microbiota. Bacterial communities indentified from different *Hydra* species (Panel modified from Fraune et al. 2009b)

to characterize the microbiota of different *Hydra* species (Fraune and Bosch 2007). When analyzing different species we discovered that they differ greatly in their associated bacterial microbiota, although they were cultured under identical conditions. Comparing the cultures maintained in the laboratory for >30 years with polyps directly isolated from the wild revealed a surprising similarity in the associated bacterial composition. The significant differences in the microbial communities between the species and the maintenance of specific microbial communities over long periods of time strongly indicate distinct selective pressures within the epithelium (Fraune and Bosch 2007, 2010) (Fig. 8.4). The diversity of the bacterial communities is comparably low and includes less than 25 bacterial phylotypes. Most of these bacteria can be cultivated (Fraune and Bosch 2007) and, therefore, are a perfect tool to decipher the contribution from each partner to the *Hydra* holobiont.

To decipher putative links between epithelial homeostasis and species-level bacterial phylotypes, we made use of mutant strain sf-1 of *Hydra magnipapillata* which has temperature sensitive interstitial stem cells (Fraune et al. 2009a). Treatment for a few hours at the restrictive temperature (28°C) induces quantitative loss of the entire interstitial cell lineage from the ectodermal epithelium, while leaving both the ectodermal and the endodermal epithelial cells undisturbed. Intriguingly, 2 weeks after temperature treatment, when the tissue was lacking not only all interstitial cells as well as nematoblasts and most nematocytes, but in addition also had a reduced number of neurons and gland cells, the bacterial composition began to change drastically. Thus, changes in epithelial homeostasis causes significant changes in the microbial community, implying a direct interaction between epithelia and microbiota.

Antimicrobial Peptides – Key Factors for Host–Bacteria Co-evolution

What is the driving force that leads to changes in microbiota composition? Because of their obvious ability to influence bacterial life, promising candidate molecules are AMPs. To investigate whether the ectopic expression of an AMP may affect the number and composition of the colonizing microbiota at the ectodermal epithelial surface, transgenic *H. vulgaris* (AEP) expressing Periculin1a in ectoderm epithelial cells were generated. Comparing the bacterial load of these transgenic polyps with wild type control polyps revealed not only a significantly lower bacterial load in transgenic polyps overexpressing Periculin1a but also, unexpectedly, drastic changes in the bacterial community structure. Analyzing the identity of the colonizing bacteria showed that the dominant β-Proteobacteria decreased in number, whereas α-Proteobacteria were more prevalent. Thus, overexpression of Periculin causes not only a decrease in the number of associated bacteria but also a changed bacterial composition (Fraune et al. 2010) (Fig. 8.5). With the transgenic polyp overexpressing periculin we have created a new Holobiont which is different from all investigated hydra species. Future efforts will be directed towards analyzing the performance of this new Holobiont phenotype under different environmental conditions. According to the hologenome theory of evolution, changing the microbial community is one relatively rapid way of adapting to novel environmental conditions.

Bacteria Deprived Holobionts Have to Suffer

The intimacy of the interaction between host and microbiota, as well as the strong evolutionary pressure (Fraune and Bosch 2007) (Fig. 8.4) to maintain a specific microbiota, points to the significance of the interkingdom association, and implies that hosts deprived of their microbiota should be handicapped. To investigate the effect of absence of microbiota in *Hydra* we recently produced gnotobiotic polyps which are devoid of any bacteria. While morphologically no differences could be observed with control polyps, *Hydra* lacking bacteria suffer from fungal infections unknown in normally cultured polyps (René Augustin, Sören Franzenburg, Julia Hahn, personal observation). Thus, the beneficial microbes associated with *Hydra* appear to produce powerful anti-fungal compounds. Additional support for a fungal defense function of the bacteria associated with *Hydra* comes from analyzing the pathogenicity of oomycetes (*Saprolegnia* spec.). Many *Saprolegnia* species cause economic and environmental damage due to their ability to infect a wide range of plants and animals (Phillips et al. 2008) causing saprolegniosis. Under standard culture conditions, *Saprolegnia* does not infect *Hydra*. Although zoospores are capable of attaching to the glycocalyx (Fig. 8.1b), germination appears to be inhibited. Under experimental conditions involving perturbations such as tissue dissociation and reaggregation in antibiotic-containing water, zoospores do not only attach, but also germinate, infect and destroy the animals completely (to be published). Thus, severe defects in the epithelial barrier and the absence of an intact microbial community facilitate germination of the spores. When different cultivable bacterial strains from hydra's microbiota were tested for their ability to inhibit germination of the spores, we identified three strains with strong inhibitory activity (to be published). Future efforts will be directed towards isolating the active substances from these bacteria which may lead to the development of novel antimycotics.

The "Green" *Hydra* – A Tripartite Interplay

In addition to uncovering mechanisms involved in host–bacteria interkingdom communication, the green species *Hydra viridissima* offers the possibility to investigate the interaction, not only between the eukaryotic host and the microbes, but also the relationship between the host, microbes

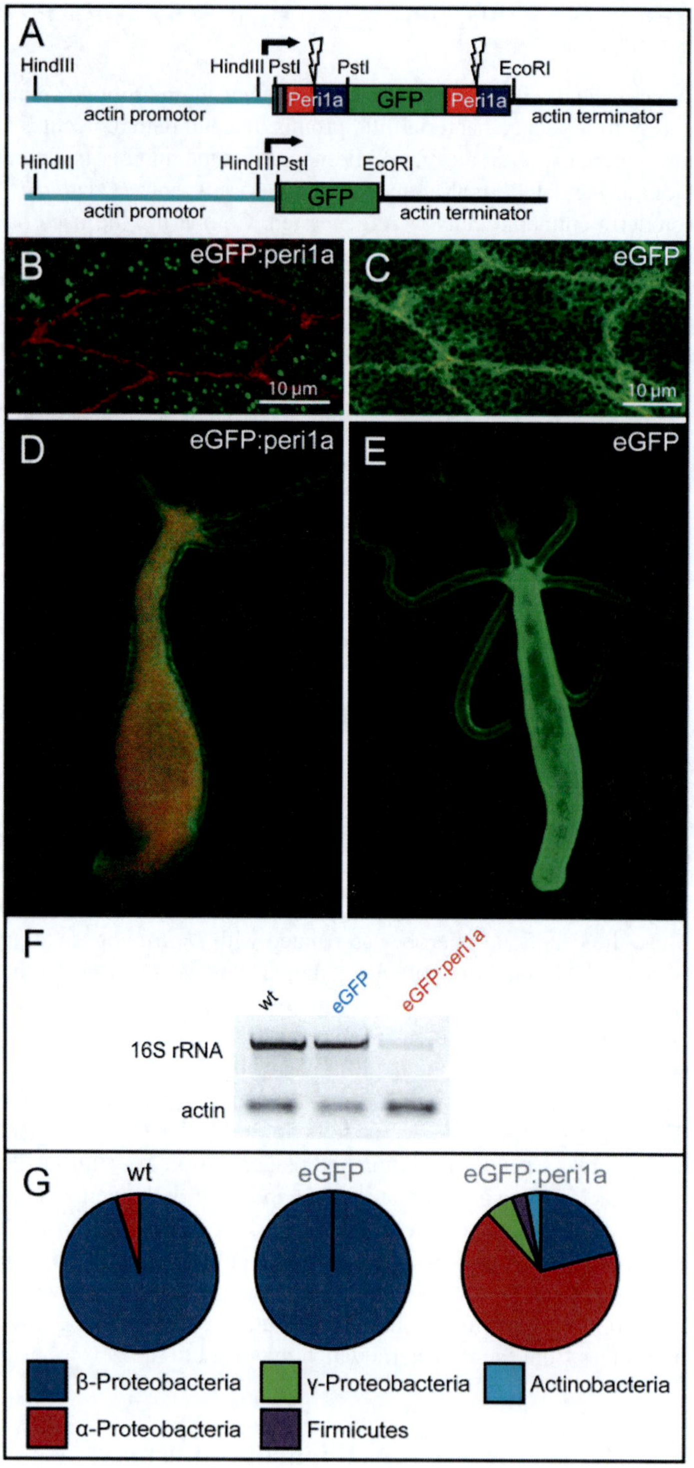

Fig. 8.5 Periculin1a controls bacterial colonization. (**a**) Expression constructs for generation of transgenic *Hydra*. (*Upper*) Construct containing periculin1a including signal peptide fused in frame at the 5′ end and periculin1a lacking signal peptide at the 3′ end of EGFP. (*Lower*) Control construct with EGFP driven by 1,386-bp actin 5′ flanking region. (**b**, **c**) Confocal micrographs of single transgenic ectodermal cells of (**b**) *Hydra vulgaris* (AEP) EGFP:periculin1a polyp (notice peptide localization in vesicles) and (**c**) *Hydra vulgaris* (AEP) EGFP control polyp (notice EGFP localization in the cytoplasm). (**d**, **e**) In vivo images of (**d**) transgenic polyp *Hydra vulgaris* (AEP) EGFP:periculin1a and (**e**) control polyp *Hydra vulgaris* (AEP) EGFP. EGFP protein is green; actin filaments are red. (**f**) PCR of genomic DNA amplifying bacterial 16S rRNA genes, equilibrated on *Hydra* actin gene. The number of bacterial 16S rRNAgenes associated with transgenic *Hydra vulgaris* (AEP) EGFP:periculin1a polyps is significantly reduced. (**g**) Comparison of bacterial composition associated with transgenic *Hydra vulgaris* (AEP) EGFP:periculin1a polyps and control (WT) polyps (Panel from Fraune et al. 2010)

and a eukaryotic symbiont (Chlorella algae). *Hydra viridissima* forms a stable symbiosis with the intracellular green algae of the *Chlorella* group (Muscatine and Lenhoff 1963) and has been a classical model system for investigating the benefits of the symbiosis between algae and *Hydra*. The symbionts are located in endodermal epithelial cells. Each alga is enclosed by an individual vacuolar membrane resembling a plastid of eukaryotic origin at an evolutionary early stage of symbiogenesis (O'Brien 1982). Proliferation of symbiont and host is tightly correlated. The photosynthetic symbionts provide nutrients to the polyps enabling *Hydra* to survive extended periods of starvation (Muscatine and Lenhoff 1963; Thorington and Margulis 1981). Symbiotic *Chlorella* is unable to grow outside the host, indicating a loss of autonomy during establishment of the intimate symbiotic interactions with *Hydra*. During sexual reproduction of the host, *Chlorella* algae are translocated into the oocyte, giving rise to a new symbiont population in the hatching embryo (Habetha et al. 2003). The molecular mechanisms involved in the recognition and/or toleration of the algal symbiont are not yet known. Currently we are taking a transcriptomic approach using microarray technology to identify hydra genes involved in the onset and control of the symbiosis. Future efforts are directed towards uncovering if and how the algal symbiont influences the bacterial community.

Conclusion and Perspectives

Here we present the fresh water polyp *Hydra* as a model to study host–microbe interaction in the context of a superorganismic organization. Molecular tools, including transgenesis as well as rich genomic and transcriptomic resources, make *Hydra* a valuable system that can be manipulated experimentally. Because of its simplicity in body structure, and the exclusive reliance on the innate immune system of epithelia, the maintenance of epithelial barriers can be investigated in the absence of the adaptive immune system and other immune-related cell types and organs. The uncovered basic molecular machinery can be related to more complex organisms, including man, since nearly all known molecules involved in innate immunity are present in *Hydra*. Moreover, due to the relatively simple microbial community of only few bacterial phylotypes (most of which can be cultured in vitro), the influence of the microbiota under healthy and disease conditions can be dissected. This might also help to decipher why bacteria under certain conditions change from commensal to pathogenic state. An evolutionary approach to innate immunity considering the co-evolution between *Hydra* and its microbial community (Fraune and Bosch 2007) may also explain why *Hydra* AMPs belong to the most powerful peptides against human pathogenic bacteria. Human associated bacteria share little or no evolutionary history with *Hydra*-associated microbes and therefore lack any adaptation to hydra peptides. Basal metazoan animals, therefore, not only provide conceptual insights into the complexity of host–microbe interactions, but may also have considerable potential as source for novel antimicrobial compounds.

Acknowledgement Research in our laboratory is supported in parts by grants from the Deutsche Forschungsgemeinschaft (DFG) and grants from the DFG Cluster of Excellence programs "The Future Ocean" and "Inflammation at Interfaces."

References

Augustin R, Siebert S, Bosch TC (2009a) Identification of a kazal-type serine protease inhibitor with potent anti-staphylococcal activity as part of Hydra's innate immune system. Dev Comp Immunol 33:830–837

Augustin R, Anton-Erxleben F, Jungnickel S, Hemmrich G, Spudy B, Podschun R, Bosch TC (2009b) Activity of the novel peptide arminin against multiresistant human pathogens shows the considerable potential of phylogenetically ancient organisms as drug sources. Antimicrob Agents Chemother 53:5245–5250

Bosch TC (2007) Why polyps regenerate and we don't: towards a cellular and molecular framework for Hydra regeneration. Dev Biol 303:421–433

Bosch TCG, David CN (1986) Immunocompetence in Hydra: epithelial cells recognize self-nonself and react against it. J Exp Biol 238:225–234

Bosch TC, Augustin R, Anton-Erxleben F, Fraune S, Hemmrich G, Zill H, Rosenstiel P, Jacobs G, Schreiber S, Leippe M, Stanisak M, Grotzinger J, Jung S, Podschun R, Bartels J, Harder J, Schroder JM (2009) Uncovering the evolutionary history of innate immunity: the simple metazoan Hydra uses epithelial cells for host defence. Dev Comp Immunol 33:559–569

Chapman JA, Kirkness EF, Simakov O, Hampson SE, Mitros T, Weinmaier T, Rattei T, Balasubramanian PG, Borman J, Busam D, Disbennett K, Pfannkoch C, Sumin N, Sutton GG, Viswanathan LD, Walenz B, Goodstein DM, Hellsten U, Kawashima T, Prochnik SE, Putnam NH, Shu S, Blumberg B, Dana CE, Gee L, Kibler DF, Law L, Lindgens D, Martinez DE, Peng J, Wigge PA, Bertulat B, Guder C, Nakamura Y, Ozbek S, Watanabe H, Khalturin K, Hemmrich G, Franke A, Augustin R, Fraune S, Hayakawa E, Hayakawa S, Hirose M, Hwang JS, Ikeo K, Nishimiya-Fujisawa C, Ogura A, Takahashi T, Steinmetz PR, Zhang X, Aufschnaiter R, Eder MK, Gorny AK, Salvenmoser W, Heimberg AM, Wheeler BM, Peterson KJ, Bottger A, Tischler P, Wolf A, Gojobori T, Remington KA, Strausberg RL, Venter JC, Technau U, Hobmayer B, Bosch TC, Holstein TW, Fujisawa T, Bode HR, David CN, Rokhsar DS, Steele RE (2010) The dynamic genome of Hydra. Nature 464:592–596

Engelberg-Kulka H, Hazan R, Amitai S (2005) mazEF: a chromosomal toxin-antitoxin module that triggers programmed cell death in bacteria. J Cell Sci 118:4327–4332

Engelberg-Kulka H, Amitai S, Kolodkin-Gal I, Hazan R (2006) Bacterial programmed cell death and multicellular behavior in bacteria. PLoS Genet 2:e135

Fraune S, Bosch TCG (2007) Long-term maintenance of species-specific bacterial microbiota in the basal metazoan Hydra. Proc Natl Acad Sci USA 104:13146–13151

Fraune S, Bosch TC (2010) Why bacteria matter in animal development and evolution. Bioessays 32:571–580

Fraune S, Abe Y, Bosch TCG (2009a) Disturbing epithelial homeostasis in the metazoan Hydra leads to drastic changes in associated microbiota. Environ Microbiol 11:2361–2369

Fraune S, Augustin R, Bosch TCG (2009b) Exploring host–microbe interactions in hydra. Microbe 4:457–462

Fraune S, Augustin R, Anton-Erxleben F, Wittlieb J, Gelhaus C, Klimovich VB, Samoilovich MP, Bosch TC (2010) In an early branching metazoan, bacterial colonization of the embryo is controlled by maternal antimicrobial peptides. Proc Natl Acad Sci USA 107:18067–18072

Habetha M, Anton-Erxleben F, Neumann K, Bosch TCG (2003) The Hydra viridis/Chlorella symbiosis. Growth and sexual differentiation in polyps without symbionts. Zoology 106:101–108

Hemmrich G, Bosch TC (2008) Compagen, a comparative genomics platform for early branching metazoan animals, reveals early origins of genes regulating stem-cell differentiation. Bioessays 30:1010–1018

Jung S, Dingley AJ, Augustin R, Anton-Erxleben F, Stanisak M, Gelhaus C, Gutsmann T, Hammer MU, Podschun R, Bonvin AM, Leippe M, Bosch TC, Grotzinger J (2009) Hydramacin-1, structure and antibacterial activity of a protein from the basal metazoan Hydra. J Biol Chem 284:1896–1905

Kortschak RD, Samuel G, Saint R, Miller DJ (2003) EST analysis of the cnidarian Acropora millepora reveals extensive gene loss and rapid sequence divergence in the model invertebrates. Curr Biol 13:2190–2195

Kusserow A, Pang K, Sturm C, Hrouda M, Lentfer J, Schmidt HA, Technau U, von Haeseler A, Hobmayer B, Martindale MQ, Holstein TW (2005) Unexpected complexity of the Wnt gene family in a sea anemone. Nature 433:156–160

Lange C, Hemmrich G, Klostermeier UC, López-Quintero JA, Miller DJ, Rahn T, Weiss Y, Bosch TC, Rosenstiel P (2011) Defining the origins of the NOD-like receptor system at the base of animal evolution. Mol Biol Evol. 28(5):1687–702. Epub 2010 Dec 23

Miller DJ, Ball EE, Technau U (2005) Cnidarians and ancestral genetic complexity in the animal kingdom. Trends Genet 21:536–539

Muscatine L, Lenhoff HM (1963) Symbiosis of Hydra with algae. J Gen Microbiol 32:6

O'Brien TL (1982) Inhibition of vacuolar membrane-fusion by intracellular symbiotic algae in Hydra-viridis (Florida strain). J Exp Zool 223:211–218

Phillips AJ, Anderson VL, Robertson EJ, Secombes CJ, van West P (2008) New insights into animal pathogenic oomycetes. Trends Microbiol 16:13–19

Putnam NH, Srivastava M, Hellsten U, Dirks B, Chapman J, Salamov A, Terry A, Shapiro H, Lindquist E, Kapitonov VV, Jurka J, Genikhovich G, Grigoriev IV, Lucas SM, Steele RE, Finnerty JR, Technau U, Martindale MQ, Rokhsar DS (2007) Sea anemone genome reveals ancestral eumetazoan gene repertoire and genomic organization. Science (New York, NY) 317:86–94

Rast JP, Smith LC, Loza-Coll M, Hibino T, Litman GW (2006) Genomic insights into the immune system of the sea urchin. Science (New York, NY) 314:952–956

Rosenberg E, Koren O, Reshef L, Efrony R, Zilber-Rosenberg I (2007) The role of microorganisms in coral health, disease and evolution. Nat Rev 5:355–362

Srivastava M, Begovic E, Chapman J, Putnam NH, Hellsten U, Kawashima T, Kuo A, Mitros T, Salamov A, Carpenter ML, Signorovitch AY, Moreno MA, Kamm K, Grimwood J, Schmutz J, Shapiro H, Grigoriev IV, Buss LW, Schierwater B, Dellaporta SL, Rokhsar DS (2008) The Trichoplax genome and the nature of placozoans. Nature 454:955–960

Technau U, Rudd S, Maxwell P, Gordon PM, Saina M, Grasso LC, Hayward DC, Sensen CW, Saint R, Holstein TW, Ball EE, Miller DJ (2005) Maintenance of ancestral complexity and non-metazoan genes in two basal cnidarians. Trends Genet 21:633–639

Thorington G, Margulis L (1981) Hydra-viridis – transfer of metabolites between Hydra and symbiotic algae. Biol Bull 160:175–188

Trembley A (1744) Mémoires, Pour Servir à l´Histoire d´un Genre de Polypes d´Eau Douce, à Bras en Frome de Cornes. Verbeek, Leiden

Wilson DS, Sober E (1989) Reviving the superorganism. J Theor Biol 136:337–356

Wittlieb J, Khalturin K, Lohmann JU, Anton-Erxleben F, Bosch TC (2006) Transgenic Hydra allow in vivo tracking of individual stem cells during morphogenesis. Proc Natl Acad Sci USA 103:6208–6211

Zilber-Rosenberg I, Rosenberg E (2008) Role of microorganisms in the evolution of animals and plants: the hologenome theory of evolution. FEMS Microbiol Rev 32:723–735

Chapter 9
Tick as a Model for the Study of a Primitive Complement System

Petr Kopacek, Ondrej Hajdusek, and Veronika Buresova

Abstract Ticks are blood feeding parasites transmitting a wide variety of pathogens to their vertebrate hosts. The transmitted pathogens apparently evolved efficient mechanisms enabling them to evade or withstand the cellular or humoral immune responses within the tick vector. Despite its importance, our knowledge of tick innate immunity still lags far beyond other well established invertebrate models, such as drosophila, horseshoe crab or mosquitoes. However, the recent release of the American deer tick, *Ixodes scapularis*, genome and feasibility of functional analysis based on RNA interference (RNAi) facilitate the development of this organism as a full-value model for deeper studies of vector-pathogen interactions.

We are focused on tick thioester-containing proteins (TEPs), which belong to the evolutionarily oldest components of the invertebrate immunity. Three phylogenetically distinct groups of TEPs are recognized in invertebrates: (1) pan-protease inhibitors of the α_2-macroglobulin type (α_2Ms), (2) C3-like complement components and (3) insect thioester-containing proteins. Previously, we have characterized a protease inhibitor of the α_2M-type from the soft tick *Ornithodoros moubata* named TAM and its ortholog from the hard tick *Ixodes ricinus* designated as IrAM. RNAi silencing of IrAM, followed by an in vitro phagocytic assay of the model tick pathogen *Chryseobacterium indologenes* by tick hemocytes revealed that IrAM is involved in this process. This result suggests a possible link between humoral and complement-like cellular immune response mediated by the interaction of an α_2M-type inhibitor with a protease originating from the invading pathogen. Our search in the *I. scapularis* genome database revealed the presence of at least nine different TEPs, comprising other α_2Ms, C3-like components and molecules related to insect TEPs. Orthologous molecules were also identified in *I. ricinus*. Thus far, *Ixodes* spp. ticks are the only organisms that seem to possess representatives of all major TEP classes. In addition to TEPs, ticks also possess other proteins putatively involved in pathogen recognition and activation of the complement-like cascade (fibrinogen related lectins or molecules related to the horseshoe crab C3 convertases).

P. Kopacek (✉) • V. Buresova
Institute of Parasitology, Biology Centre, Academy of Sciences of the Czech Republic, Branisovská 31, Ceské Budejovice CZ-370 05, Czech Republic
e-mail: kopajz@paru.cas.cz

O. Hajdusek
Institute of Parasitology, Biology Centre, Academy of Sciences of the Czech Republic, Branisovská 31, Ceské Budejovice CZ-370 05, Czech Republic

Institut de Biologie Moléculaire et Cellulaire (IBMC), Université Louis Pasteur, 15, rue René Descartes, 67084 Strasbourg Cedex, France

E. Mylonakis et al. (eds.), *Recent Advances on Model Hosts*,
Advances in Experimental Medicine and Biology 710, DOI 10.1007/978-1-4419-5638-5_9,

Taken together, functional genomics in ticks using RNAi promise to significantly extend our knowledge about the evolution and function of the primordial complement system and potentially help to combat transmission of tick-borne pathogens.

Ticks as Vectors of Tick-Borne Diseases

Ticks are blood-sucking ectoparasites belonging to the Phylum Arthropoda, Class Arachnida, Subclass Acari, Order Parasitoformes and Suborder Ixodida (Nava et al. 2009). More than the 900 tick species described until now are divided into two major families – Argasidae (also known as soft ticks) and Ixodidae (generally referred to as hard ticks) (Barker and Murell 2008). Soft ticks with a characteristic leathery cuticle (e.g. the genera *Argas* or *Ornithodoros*) are multi-host parasites, having several nymphal stages (2–8) that all feed rapidly on their hosts (within minutes to hours). The adult ticks can feed repeatedly and the females deposit a few hundreds of eggs after each feeding. The life span of Argasidae is quite long (up to several years), and nymphs or adults are capable to live between individual blood meals for months. The hard ticks, denoted as such according to its scutum that is a sclerotized surface plate, have only three parasitic stages, larvae, nymphs and adults, which feed slowly for several days. Most hard ticks feed on different hosts (three-host ticks) as e.g. the species of the genera *Ixodes, Amblyomma, Dermacentor, and Haemaphysalis*. Other species remain and feed on the same host, as is the case of a veterinary important cattle tick *Rhipicephalus* (formerly *Boophilus*) *microplus*. The fully engorged adult females drop off the host, digest the blood meal within several weeks, lay thousands of eggs on the ground and die.

While massive tick infestation may lead to severe blood loss, the main danger of ticks to their hosts is their capability to transmit a wide variety of pathogens, comprising of viruses, bacteria and protozoa (Jongejan and Uilenberg 2004), causing serious diseases in humans and domestic animals (de la Fuente et al. 2008a). The extended lifespan, long-lasting blood feeding, modulation of the host haemostatic, inflammatory and immune responses by the broad spectrum of molecules present in tick saliva (Francischetti et al. 2009), as well as the absence of proteases in gut lumen (because ticks digest blood intracellularly) (Sonenshine 1991), make the tick an exceptionally friendly environment for pathogen proliferation and transmission. Therefore, ticks have to possess efficient defense mechanisms to eliminate or at least mitigate microbial infections. On the other hand, during a long-term co-evolution, the transmitted pathogens have apparently adapted to evade or withstand the cellular and humoral immune responses within the vector. Despite its importance, we still know very little about the interactions at the tick-pathogen interface.

During the past two decades, our knowledge of innate immunity in invertebrates has developed rapidly thanks to the considerable body of work carried out on the model insect *Drosophila melanogaster* (Ferrandon et al. 2007), and other arthropods, such as the horseshoe crab, freshwater crayfish or ascidia prochordates (Iwanaga and Lee 2005). A remarkable progress was also achieved in investigation of the immune responses and parasite transmission in insect disease vectors, especially mosquitoes (Osta et al. 2004) and tsetse flies (Lehane et al. 2004).

In contrast to versatile models like drosophila, allowing easy genetic manipulation, or horseshoe crab with its peerless size for biochemical studies, ticks are small, have a complex and long life cycle, plus are rather difficult to rear in the laboratory and manipulate for biological experiments. Therefore, only a limited number of cellular immune reactions and sufficiently characterized immune molecules has been described for a variety of tick species (Sonenshine and Hynes 2008; Kopacek et al. 2010), which allowed only careful interpretation based on the immune system of other model arthropods. A few years ago, it seemed that research on the tick immune system would remain forever at a purely comparative level. However, several recent advances reversed this pessimistic outlook. The availability of large scale EST data sets on different tick species, the first draft of the American deer tick *Ixodes scapularis* genome that was released in December 2008 (http://www.vectorbase.org) (Nene 2009), and the feasibility of RNA interference (RNAi) in ticks (de la Fuente et al. 2007) set the framework for exploiting functional genomics in dissecting the mechanisms underlying tick-host-pathogen interactions. Several examples showing successful application of RNAi to reveal the role that certain molecules play at the interface between ticks and transmitted pathogens have been reviewed recently (de la Fuente et al. 2008b). Naturally, the most focus has been given to the spirochetes of the genus *Borrelia*, the causative agents of the Lyme disease and tick-borne relapsing fever (de Silva et al. 2009). During blood feeding, borrelia spirochetes multiply in the tick gut lumen and change the expression pattern of their outer surface proteins, facilitating their exit from the gut, dissemination in the hemolymph, migration and entrance to the salivary glands, which are steps needed ultimately for infection of the mammalian host (Hovius et al. 2007). Also transmission of other important pathogens like the intracellular rickettsial bacteria of the genus *Anaplasma* (Kocan et al. 2008) or the apicomplexan protozoa *Babesia sp.* (Chauvin et al. 2009) and *Theileria sp.* (Bishop et al. 2008) proceeds from initial invasion of the midgut cells to infection of other tick tissues including the salivary glands. In all cases, the microbes have to overcome the barrier of cellular and humoral immune responses within the tick hemolymph, which in addition to interactions within the gut and salivary glands, represents a critical part of the whole tick-pathogen interface.

Immune Reactions in the Tick Hemolymph

Based on morphological and physiological studies of hemolymph cells in different tick species, three basic classes of tick hemocytes, namely plasmatocytes, granulocytes I and granulocytes II have been described (Kuhn and Haug 1994; Borovickova and Hypsa 2005), out of which the first two types are capable of phagocytosis of foreign material and invading microbes (Inoue et al. 2001; Loosova et al. 2001; Pereira et al. 2001). Interestingly, engulfment of the Lyme disease agent *Borrelia burgdorferi* by tick hemocytes was shown to be performed by "coiling" phagocytosis, a manner similar to that performed by vertebrate phagocytic cells (Rittig et al. 1996). Coleman et al. demonstrated that *B. burgdorferi* use plasmin from the host blood bound on their surface to overcome the midgut barrier, disseminate into the tick hemolymph and migrate towards the salivary glands. At least some of these spirochetes were phagocytosed by tick hemocytes during this process (Coleman et al. 1997). Although no cell line originating from tick hemocytes has been available to the date, some tick cell lines derived from embryonic cells (Bell-Sakyi et al. 2007) display remarkable phagocytic activity and have been used as useful models for deeper insight into tick cell-pathogen interaction (Mattila et al. 2007; Blouin et al. 2002).

In our laboratory, we have observed the remarkable phagocytic activity of tick hemocytes against Gram$^-$ *Chryseobacterium indologenes* (Buresova et al. 2006). This yellow-pigmented bacterium, which is resistant to common antibiotics, turned out to be responsible for the ultimate extinction of our colony of the soft ticks *Ornithodoros moubata*. Prior to this unlucky event, the *O. moubata* ticks were used for classical biochemical studies on tick hemolymph, since they allowed us to collect enough starting material for protein purifications. We have characterized two molecules from the hemolymph of this species having a potential in tick immunity, namely a lectin Dorin M (Kovar et al. 2000) and the universal protease inhibitor of α_2-macroglobulin type designated as TAM (Kopacek et al. 2000).

TAM Is the First Thioester-Containing Protein Described in Ticks

The proteins of the α_2-macroglobulin family, more aptly referred to as thioester-containing proteins (TEPs), belong to the evolutionarily oldest and best conserved components of innate immunity. In invertebrates, this family comprises three phylogenetically distinct groups: (1) α_2-macroglobulins

(α_2Ms) – the universal macromolecular protease inhibitors; (2) C3/C4/C5 components of the complement system; (3) insect thioester-containing proteins (TEPs) (Blandin and Levashina 2004). The pan-protease inhibitors of α_2M class are abundantly present in plasma of vertebrates and were described also in a number of mainly aquatic invertebrates (Dodds and Day 1996), but not yet in insects. They are believed to have a role in immunity by guarding against the undesired action of proteases of foreign origin, including those from invading pathogens (Armstrong 2010). The mammalian complement activation cascades (classical, alternative and lectin pathway) merge in the proteolytic activation of the central component C3 (Ricklin et al. 2010). The presence of this molecule could be tracked back in the evolution to the most primitive invertebrates, indicating that primitive complement system could exist for more than 1 billion years (Zhu et al. 2005; Endo et al. 2006; Nonaka and Kimura 2006). A complement-like function has been demonstrated also for insect TEPs. TEP1 from the malaria vector, the mosquito *Anopheles gambiae*, binds to the surface of bacteria and promotes their phagocytosis (Levashina et al. 2001). Moreover, this molecule was proven to be the determinant of the mosquito vectorial capacity, since it binds to and mediates killing of *Plasmodium* ookinetes in the midgut (Blandin et al. 2004, Blandin et al. 2008).

The proteins of the TEP family are synthesized as large precursors of about 1,500 amino-acid residues and arranged in a similar array of eight macroglobulin domains (Janssen et al. 2005; Baxter et al. 2007a; Doan and Gettins 2007). They become functional upon proteolytic cleavage of the highly variable segment occurring within the central macroglobulin six domain, referred to as the 'bait' region, anaphylatoxin domain or protease sensitive region in α_2Ms, C3/C4/C5 complement components or insect TEPs, respectively (Blandin et al. 2008). Proteolytic cleavage causes a rapid conformational change leading to the spatial exposure of a domain containing a highly reactive thioester (TE) bond. Activated TE bond of C3 or insect TEP molecules covalently binds to the microbial surface facilitating their phagocytosis or firmly captures the protease entrapped inside the macromolecular cage of α_2M inhibitors. Although the TE bond is a hallmark for TEPs, some members of the family, like the C5 complement component or several insect TEPs, lack the conserved CGEQ motif but are assigned to this protein family by sequence and/or structural similarities.

TAM, the α_2M isolated and sequenced from the soft tick *O. moubata* (Kopacek et al. 2000; Saravanan et al. 2003) was the first representative of the TEP family described in ticks. The protein is abundant in the tick plasma and is capable to inhibit proteases like trypsin or thermolysin by an entrapping mechanism typical for other α_2Ms. TAM has a mass of about 420 kDa and is composed of two identical non-covalently bound subunits, which are post-translationally cleaved into two chains in a fashion resembling the processing of the α- and β-chains of C3 and C4 complement components. Interestingly, the 'bait' region of TAM is considerably diversified by alternative splicing (Saravanan et al. 2003). An ortholog of TAM was identified and characterized by a reverse genetic approach in the hard tick *Ixodes ricinus* and named IrAM (Buresova et al. 2009). In addition to high sequence homology, TAM and IrAM share also a similar pattern of disulfide bridges, glycosylation sites and variability within the 'bait' region generated by alternative splicing (Fig. 9.1).

IrAM Is Involved in Phagocytosis of *Chryseobacterium indologenes*

Based on our previous investigation of *C. indologenes* pathogenesis in *O. moubata* and *I. ricinus* (Buresova et al. 2006), we were able to establish a robust in vitro phagocytic assay linked with a preceding hemolymph pre-treatment or RNAi knockdown of a selected gene, as schematically depicted in Fig. 9.2. Using this assay, we have found that silencing of IrAM, which results in its complete depletion from the tick hemolymph, significantly decreases the phagocytosis of *C. indologenes* by tick hemocytes (Buresova et al. 2009). An even more profound effect was achieved

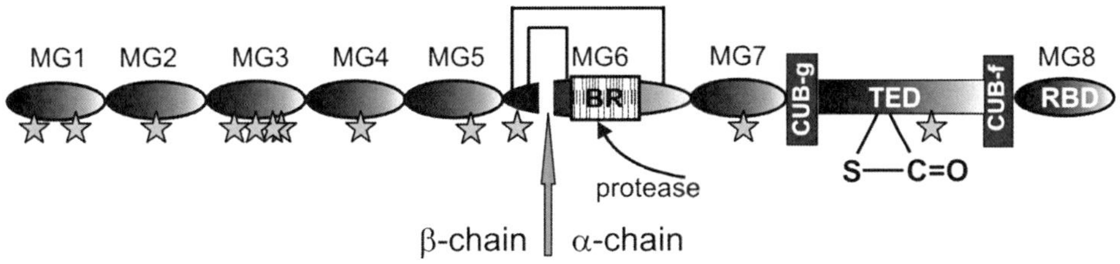

Fig. 9.1 Predicted structure of tick α_2-macroglobulins: TAM from the soft tick *Ornithodoros moubata* and IrAM1 from the hard tick *Ixodes ricinus*. The domain arrangement of TAM and IrAM1 were predicted in comparison to the structures of human C3 (Janssen et al. 2005), mosquito TEP1 (Baxter et al. 2007b) and human α_2M (Doan and Gettins 2007). The TEP proteins consist of eight macroglobulin domains (*MG1*-8) that form two major rings named the β- and α-chain. Unlike the majority of known α_2M, the tick orthologs are post-translationally cleaved during secretion in a manner resembling the processing of C3 and C4 complement components (indicated by the grey arrow). The β-chain consisting of MG1-5 and part of the MG6 domain is heavily glycosylated (gray stars) and remains linked to the α-chain by two disulfide bridges. *BR* 'bait' region, a highly variable segment that serves as a target for a broad spectrum of proteases. The 'bait' region in tick α_2Ms is further modified by an alternative splicing. Two cubilins (*CUB*) domains delimit the conserved thioester domain (*TED*), containing the highly reactive thioester bond that can be deactivated by small primary amines (e.g. methylamine). The MG8 segment contains the receptor binding domain (RBD), which assists in scavenging the reacted α_2M-protease complex from circulation (Adapted from Buresova et al. 2009)

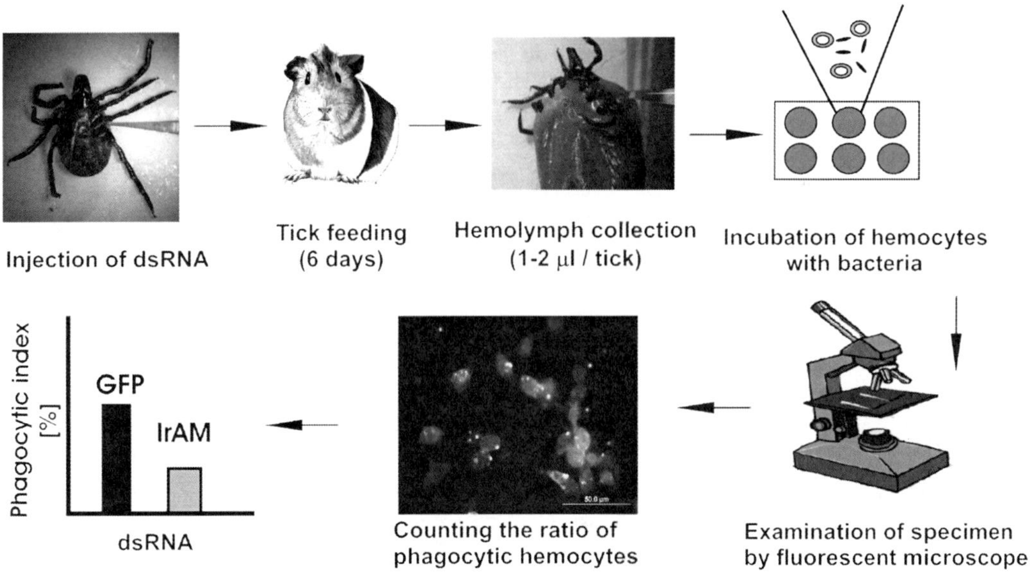

Fig. 9.2 Flow chart of the RNA interference linked with phagocytic assay. The adult *I. ricinus* females are injected with gene-specific dsRNA (IrAM) or green fluorescent protein (*GFP*) dsRNA as a control. One day later, ticks are allowed to feed naturally on the host. The hemolymph is collected from semi-engorged tick (day 6 of feeding), appropriately diluted in physiological buffer and incubated with a defined number of bacteria (*Chryseobacterium indologenes*). The bacteria are visualized by specific antibodies and the ratio of phagocytic hemocytes engulfing the bacteria to non-phagocytic hemocytes (phagocytic index) is calculated

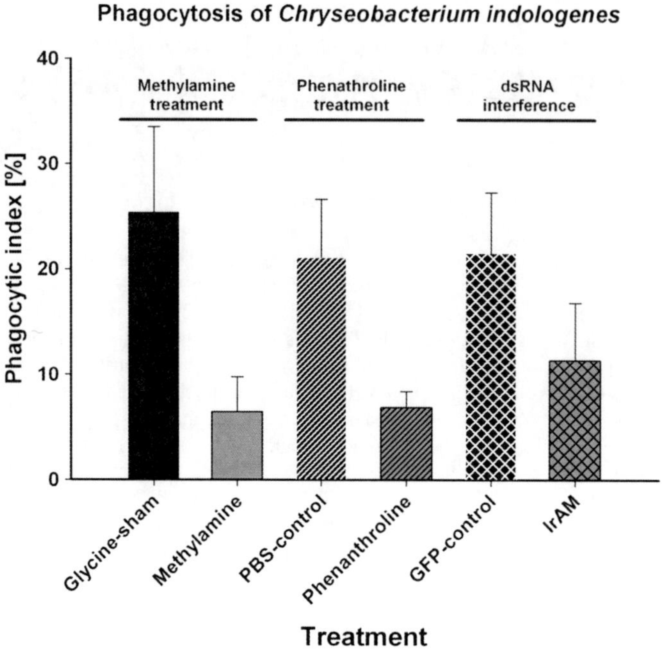

Fig. 9.3 The effect of methylamine treatment, protease inhibition and RNAi silencing of tick α_2-macroglobulin (*IrAM*) on the phagocytosis of *Chryseobacterium indologens* by tick hemocytes. **Methylamine treatment** – Pre-treatment of *I. ricinus* hemolymph with methylamine, which deactivates the thioester bonds present in TEPs, markedly reduces phagocytosis of *C. indologenes* by tick hemocytes. The same concentration of L-glycine was used in parallel as a sham. **Phenathroline treatment** – The pre-incubation of *C. indologenes* with 1,10-phenthroline, a specific Zn^{2+}-metalloprotease inhibitor, has an adverse affect on bacterium phagocytosis. **dsRNA interference** – RNAi knockdown of *I. ricinus* α_2-macroglobulin (*IrAM*) significantly decreases phagocytosis of *C. indologenes*. Injection of GFP dsRNA is used as a control. These data suggest that a cellular response against *C. indologenes* involves an interaction of an α_2M–type inhibitor with a protease secreted by the invading pathogen (Adapted from Buresova et al. 2009)

upon pre-treatment of tick plasma with methylamine, which deactivates the thioester bound in TEPs. This result suggested that there are possibly other TEPs beside IrAM involved in this process (Buresova et al. 2009). We have also found that phagocytosis of *C. indologenes* is highly reduced in the presence of 1,10-phenantroline, an inhibitor that specifically blocks the activity of a Zn^{2+}-dependent metalloprotease secreted by the *C. indologenes* (Buresova et al. 2009). These data, summarized in Fig. 9.3, indicate that phagocytosis of *C. indologenes* is mediated via interaction of IrAM with a protease originating from the invading pathogen. By contrast, the phagocytosis of the agent of Lyme disease *B. burgorferi* was not affected by methylamine and phenathroline treatments as well as IrAM silencing. This observation suggests that engulfment of the spirochetes by "coiling" phagocytosis is mediated by a different mechanism that is independent of the TEPs (Buresova et al. 2009).

The *Ixodes* sp. Hard Ticks Possess Nine Different TEPs

In order to answer the question how many thioester proteins are in *Ixodes* sp. ticks, we searched for TEPs in the genome of the representative hard tick *I. scapularis*. All together, we have identified genes for nine proteins belonging to the TEP family and tentatively named them as IsAM1-9 (Buresova 2009). A blast search through the available databases revealed that three of these proteins, termed IsAM1, 2 and 4, belong to the α_2-macroglobulins. IsAM3 is clearly related to the insect TEPs while IsAM8 and 9 display the highest similarity with insect TEPs recently renamed macroglobulin-complement-related (MCRs) (Stroschein-Stevenson et al. 2006). Finally, IsAM5-7 appear to belong to the category of C3-like complement components. To our knowledge, no other invertebrate organism possessing representatives of all major TEP groups has been described. This finding promotes tick as an exceptional model for further research on the role of TEPs in innate immunity and the origin of the complement system in general.

We were able to identify all nine IsAM1-9 orthologs in *I. ricinus* and designated them as IrAM1-9. The sequences of corresponding TEP orthologs are almost identical which allows to perform functional genomic studies valid for both *Ixodes* species. Most IrAMs are expressed mainly in hemocytes and salivary glands except for IrAM 3 (the insect TEP-like protein), which seems to be expressed exclusively in the salivary glands (Buresova 2009).

It has been demonstrated for mosquito hemocytes (Moita et al. 2005) and drosophila S2 cells (Stroschein-Stevenson et al. 2006) that phagocytosis of Gram$^-$, Gram$^+$ bacteria and yeast depends on different TEPs involved in distinct complement-like pathways. Our preliminary results indicated that phagocytosis of *E. coli* and the yeast *Candida albicans* is also reduced upon hemolymph treatment with methylamine, suggesting TEP involvement. It remains to be determined which out of the nine IrAM(s) are involved in the phagocytosis of these microbes. Certainly, the most interesting results should bring carefully selected combinations of individual IrAMs silencing vs. phagocytosis of the relevant tick-borne pathogens.

Ficolin-Like Lectins and Other Putative Components of a Primitive Complement System in Ticks

Tick lectins/hemagglutinins are our long-term focus since the specific protein-carbohydrate interactions apparently play a key role in self/nonself recognition (Grubhoffer et al. 2008). Dorin M was the first lectin purified and characterized from any tick species. It has a binding specificity to N-acetyl-D-hexosamines, sialic acid and sialo-glycoconjugates like fetuin (Kovar et al. 2000). Native Dorin M has a mass of about 640 kDa and is composed of non-covalently bond 37 kDa subunits. The protein has three glycosylation sites, bearing two typical high-mannose-type glycans and one truncated high-mannose-type modified with fucose bound to the core N-acetylglycosamine (Man et al. 2008). Dorin M is mainly expressed in tick hemocytes and salivary glands. The C-terminal part of Dorin M is homologous to the γ-chain of vertebrate fibrinogens and therefore belongs to the family of fibrinogen-related proteins (FREPs). Dorin M is, together with Ixoderin A and B identified in the hard tick *I. ricinus* (Rego et al. 2005), are the most closely related to the tachylectins 5A and 5B from the horseshoe crab *Tachypleus tridentatus* (Gokudan et al. 1999), which function as pattern recognition molecules (Kawabata and Tsuda 2002). These molecules are also related to mammalian ficolins (Rego et al. 2006), however, by contrast, they all lack the N-terminal collagen-like domain (Fig. 9.4). A complement-related function was demonstrated for the tachylectin 5A, B orthologs named carcinolectins CL5a and CL5b from the horseshoe crab *Carcinoscorpius rotundicauda* (Zhu et al. 2005). The authors reported that carcinolectins CL5a and CL5b bind together with a complement C3-like molecule onto the surface of different microbes, such as Gram$^+$- and Gram$^-$ -bacteria as well as fungi (Zhu et al. 2005). Additionally, they identified a molecule named CrC2/Bf (Zhu et al. 2005), which is homologous to the vertebrate C3-activating convertases, namely the C2 complement component acting in the lectin and classical pathway or the factor B (Bf) involved in the alternative pathway of complement activation. A molecule with a similar protein architecture, composed of several repeats of complement control protein modules (CCP) followed by a von Willebrand factor type A domain and C-terminal trypsin-type serine protease can

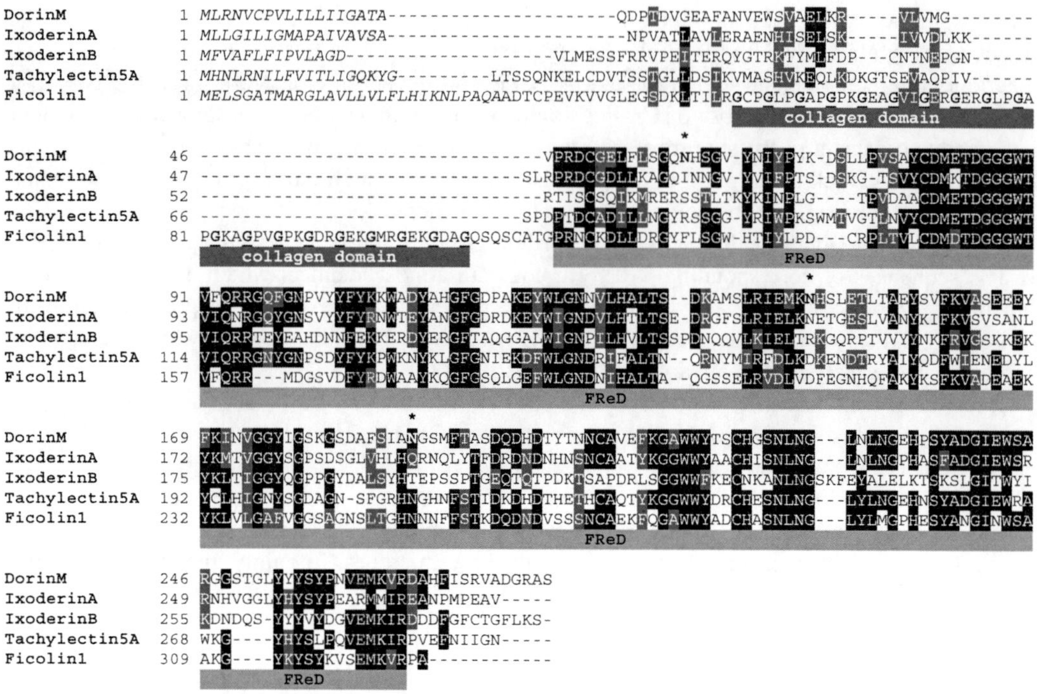

Fig. 9.4 Multiple sequence alignment of the Dorin M and other tick FREPs with tachylectin 5A and human ficolin. Dorin M – *Ornithodoros moubata* plasma lectin (GenBank AY333989); IxoderinA, IxoderinB – fibrinogen related proteins identified in the hard tick *Ixodes ricinus* (GenBank AY341424 and AY643518, respectively); Tachylectin 5A – horsehoe crab *Tachypleus tridentatus* lectin (GenBank AB02737). Ficolin1 – human ficolin (GenBank NM_002003). The predicted signal peptides are in italics. The glycine repeats in the collagen domain of human ficolin are bold-underlined. The glycosylated sites of Dorin M are marked with asterisks. FReD – fibrinogen related domain (Adapted from Rego et al. 2005; Rego et al. 2006)

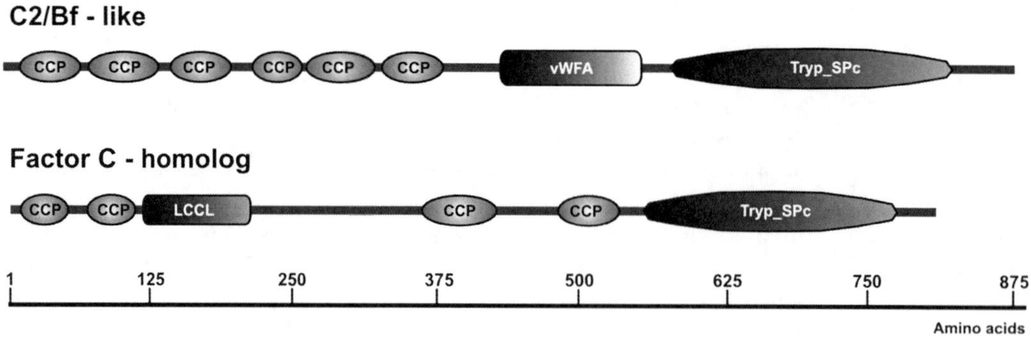

Fig. 9.5 The domain arrangement of putative C3-convertases identified in the genome of the hard tick *Ixodes scapularis*. The domain arrangement were achieved by the protein blast search using the NCBI server (http://blast.ncbi.nlm.nih.gov/Blast.cgi) using the Vector base genes ISCW023729 and ISCW002489 for C2/Bf-like and Factor C – homolog, respectively. *CCP* Complement control protein module, *vWFA* von Willebrand factor type A domain, *Tryp_SPc* trypsin-0like serine protease, *LCCL* LCCL superfamily domain

be also found in the tick genome (Fig. 9.5). In addition, ticks also possess a homolog of the horseshoe crab factor C (Fig. 9.5), which recognizes pathogen-associated molecular patterns, such as lipopolysaccharides on the bacterial surface, and triggers a coagulation cascade in the horseshoe crab hemolymph (Kawabata et al. 2009). Recently, it was demonstrated that factor C also acts as a LPS-responsive C3 convertase on the surface of invading Gram^{-}-bacteria in the initial phase of horseshoe crab complement activation (Ariki et al. 2008).

All together, these data indicate that ticks possess components of a primitive complement-like system similar to that reported in horseshoe crabs (Zhu et al. 2005; Ariki et al. 2008). Therefore, further functional genomics studies in ticks promise to become a valuable counterpart to the biochemical and functional studies in other organisms, and could significantly contribute to our understanding of the evolution of complement system and its role in innate immunity in general.

Acknowledgements This work was supported by grant P506/110/2136 to P.K. from the Grant Agency of the Czech Republic, the Research Centre LC06009 and Research projects Z60220518 and MSMT6007665801 from Ministry of Education, Youth, and Sports of the Czech Republic.

References

Ariki S, Takahara S, Shibata T et al (2008) Factor C acts as a lipopolysaccharide-responsive C3 convertase in horseshoe crab complement activation. J Immunol 181:7994–8001

Armstrong PB (2010) Role of α2-macroglobulin in the immune response of invertebrates. Invert Surviv J 7:165–180

Barker SC, Murell A (2008) Systematics and evolution of ticks with a list of valid genus and species names. In: Bowman AS, Nuttall PA (eds) Ticks: biology, disease and control. Cambridge University Press, Cambridge/New York, pp 1–39

Baxter RH, Chang CI, Chelliah Y et al (2007) Structural basis for conserved complement factor-like function in the antimalarial protein TEP1. Proc Natl Acad Sci USA 104:11615–11620

Bell-Sakyi L, Zweygarth E, Blouin EF et al (2007) Tick cell lines: tools for tick and tick-borne disease research. Trends Parasitol 23:450–457

Bishop R, Musoke A, Skilton R (2008) *Theileria*: life cycle stages associated with the ixodid tick vector. In: Bowman AS, Nuttall PA (eds) Ticks: biology, disease and control. Cambridge University Press, New York, pp 308–324

Blandin S, Levashina EA (2004) Thioester-containing proteins and insect immunity. Mol Immunol 40:903–908

Blandin S, Shiao SH, Moita LF et al (2004) Complement-like protein TEP1 is a determinant of vectorial capacity in the malaria vector *Anopheles gambiae*. Cell 116:661–670

Blandin SA, Marois E, Levashina EA (2008) Antimalarial responses in *Anopheles gambiae*: from a complement-like protein to a complement-like pathway. Cell Host Microbe 3:364–374

Blouin EF, de la Fuente J, Garcia-Garcia JC et al (2002) Applications of a cell culture system for studying the interaction of *Anaplasma marginale* with tick cells. Anim Health Res Rev 3:57–68

Borovickova B, Hypsa V (2005) Ontogeny of tick hemocytes: a comparative analysis of *Ixodes ricinus* and *Ornithodoros moubata*. Exp Appl Acarol 35:317–333

Buresova V (2009) Function of the α2-macroglobulin protein family in the immune response of the tick *Ixodes ricinus* [Ph.D.]. Ceske Budejovice. Faculty of Science, University of South Bohemia

Buresova V, Franta Z, Kopacek P (2006) A comparison of *Chryseobacterium indologenes* pathogenicity to the soft tick *Ornithodoros moubata* and hard tick *Ixodes ricinus*. J Invertebr Pathol 93:96–104

Buresova V, Hajdusek O, Franta Z et al (2009) IrAM-An alpha2-macroglobulin from the hard tick *Ixodes ricinus*: characterization and function in phagocytosis of a potential pathogen *Chryseobacterium indologenes*. Dev Comp Immunol 33:489–498

Chauvin A, Moreau E, Bonnet S et al (2009) *Babesia* and its hosts: adaptation to long-lasting interactions as a way to achieve efficient transmission. Vet Res 40:37

Coleman JL, Gebbia JA, Piesman J (1997) Plasminogen is required for efficient dissemination of *B. burgdorferi* in ticks and for enhancement of spirochetemia in mice. Cell 89:1111–1119

de la Fuente J, Kocan KM, Almazan C et al (2007) RNA interference for the study and genetic manipulation of ticks. Trends Parasitol 23:427–433

de la Fuente J, Estrada-Pena A, Venzal JM et al (2008a) Overview: ticks as vectors of pathogens that cause disease in humans and animals. Front Biosci 13:6938–6946 de la Fuente J, Kocan KM, Almazan C et al (2008b) Targeting the tick-pathogen interface for novel control strategies. Front Biosci 13:6947–6956

de Silva AM, Tyson KR, Pal U (2009) Molecular characterization of the tick-*Borrelia* interface. Front Biosci 14:3051–3063

Doan N, Gettins PG (2007) Human alpha2-macroglobulin is composed of multiple domains, as predicted by homology with complement component C3. Biochem J 407:23–30

Dodds AW, Day AJ (1996) Complement-related proteins in invertebrates. In: Soderhall K, Iwanaga S, Vasta GR (eds) New directions in invertebrate immunology. SOS Publications, Fair Haven, pp 303–342

Endo Y, Takahashi M, Fujita T (2006) Lectin complement system and pattern recognition. Immunobiology 211:283–293

Ferrandon D, Imler JL, Hetru C et al (2007) The *Drosophila* systemic immune response: sensing and signalling during bacterial and fungal infections. Nat Rev Immunol 7:862–874

Francischetti IM, Sa-Nunes A, Mans BJ et al (2009) The role of saliva in tick feeding. Front Biosci 14:2051–2088

Gokudan S, Muta T, Tsuda R et al (1999) Horseshoe crab acetyl group-recognizing lectins involved in innate immunity are structurally related to fibrinogen. Proc Natl Acad Sci USA 96:10086–10091

Grubhoffer L, Rego ROM, Hajdušek O et al (2008) Tick lectins and fibrinogen-related proteins. In: Bowman AS, Nuttall PA (eds) Ticks: biology, disease and control. Cambridge University Press, Cambridge/New York, pp 127–142

Hovius JW, van Dam AP, Fikrig E (2007) Tick-host-pathogen interactions in Lyme borreliosis. Trends Parasitol 23:434–438

Inoue N, Hanada K, Tsuji N et al (2001) Characterization of phagocytic hemocytes in *Ornithodoros moubata* (Acari: Ixodidae). J Med Entomol 38:514–519

Iwanaga S, Lee BL (2005) Recent advances in the innate immunity of invertebrate animals. J Biochem Mol Biol 38:128–150

Janssen BJ, Huizinga EG, Raaijmakers HC et al (2005) Structures of complement component C3 provide insights into the function and evolution of immunity. Nature 437:505–511

Jongejan F, Uilenberg G (2004) The global importance of ticks. Parasitology 129(Suppl):S3–14

Kawabata S, Tsuda R (2002) Molecular basis of non-self recognition by the horseshoe crab tachylectins. Biochim Biophys Acta 1572:414–421

Kawabata S, Koshiba T, Shibata T (2009) The lipopolysaccharide-activated innate immune response network of the horseshoe crab. Invert Surviv J 6:59–77

Kocan KM, de la Fuente J, Blouin EF (2008) Advances toward understanding the molecular biology of the *Anaplasma*-tick interface. Front Biosci 13:7032–7045

Kopacek P, Weise C, Saravanan T et al (2000) Characterization of an alpha-macroglobulin-like glycoprotein isolated from the plasma of the soft tick *Ornithodoros moubata*. Eur J Biochem 267:465–475

Kopacek P, Hajdusek O, Buresova V et al (2010) Tick innate immunity. In: Soderhall K (ed) Invertebrate immunity. Landes Bioscience and Springer Science+Business Media, New York, pp 137–162

Kovar V, Kopacek P, Grubhoffer L (2000) Isolation and characterization of Dorin M, a lectin from plasma of the soft tick *Ornithodoros moubata*. Insect Biochem Mol Biol 30:195–205

Kuhn KH, Haug T (1994) Ultrastructural, cytochemical, and immunocytochemical characterization of hemocytes of the hard tick *Ixodes ricinus* (Acari Chelicerata). Cell Tissue Res 277:493–504

Lehane MJ, Aksoy S, Levashina E (2004) Immune responses and parasite transmission in blood-feeding insects. Trends Parasitol 20:433–439

Levashina EA, Moita LF, Blandin S et al (2001) Conserved role of a complement-like protein in phagocytosis revealed by dsRNA knockout in cultured cells of the mosquito, *Anopheles gambiae*. Cell 104:709–718

Loosova G, Jindrak L, Kopacek P (2001) Mortality caused by experimental infection with the yeast *Candida haemulonii* in the adults of *Ornithodoros moubata* (Acarina: Argasidae). Folia Parasitol (Praha) 48:149–153

Man P, Kovar V, Sterba J et al (2008) Deciphering Dorin M glycosylation by mass spectrometry. Eur J Mass Spectrom (Chichester, Eng) 14:345–354

Mattila JT, Munderloh UG, Kurtti TJ (2007) Phagocytosis of the Lyme disease spirochete, *Borrelia burgdorferi*, by cells from the ticks, *Ixodes scapularis* and *Dermacentor andersoni*, infected with an endosymbiont, *Rickettsia peacockii*. J Insect Sci 7(58):1–12

Moita LF, Wang-Sattler R, Michel K et al (2005) In vivo identification of novel regulators and conserved pathways of phagocytosis in *A. gambiae*. Immunity 23:65–73

Nava S, Guglielmone AA, Mangold AJ (2009) An overview of systematics and evolution of ticks. Front Biosci 14:2857–2877

Nene V (2009) Tick genomics–coming of age. Front Biosci 14:2666–2673

Nonaka M, Kimura A (2006) Genomic view of the evolution of the complement system. Immunogenetics 58:701–713

Osta MA, Christophides GK, Vlachou D et al (2004) Innate immunity in the malaria vector *Anopheles gambiae*: comparative and functional genomics. J Exp Biol 207:2551–2563

Pereira LS, Oliveira PL, Barja-Fidalgo C et al (2001) Production of reactive oxygen species by hemocytes from the cattle tick *Boophilus microplus*. Exp Parasitol 99:66–72

Rego RO, Hajdusek O, Kovar V et al (2005) Molecular cloning and comparative analysis of fibrinogen-related proteins from the soft tick *Ornithodoros moubata* and the hard tick *Ixodes ricinus*. Insect Biochem Mol Biol 35:991–1004

Rego RO, Kovar V, Kopacek P et al (2006) The tick plasma lectin, Dorin M, is a fibrinogen-related molecule. Insect Biochem Mol Biol 36:291–299

Ricklin D, Hajishengallis G, Yang K et al (2010) Complement: a key system for immune surveillance and homeostasis. Nat Immunol 11:785–797

Rittig MG, Kuhn KH, Dechant CA et al (1996) Phagocytes from both vertebrate and invertebrate species use "coiling" phagocytosis. Dev Comp Immunol 20:393–406

Saravanan T, Weise C, Sojka D et al (2003) Molecular cloning, structure and bait region splice variants of alpha2-macroglobulin from the soft tick *Ornithodoros moubata*. Insect Biochem Mol Biol 33:841–851

Sonenshine DE (1991) Biology of ticks, vol 1. Oxford University Press, New York

Sonenshine DE, Hynes WL (2008) Molecular characterization and related aspects of the innate immune response in ticks. Front Biosci 13:7046–7063

Stroschein-Stevenson SL, Foley E, O'Farrell PH et al (2006) Identification of *Drosophila* gene products required for phagocytosis of *Candida albicans*. PLoS Biol 4:e4

Zhu Y, Thangamani S, Ho B et al (2005) The ancient origin of the complement system. EMBO J 24:382–394

Chapter 10
Models Hosts for the Study of Oral Candidiasis

Juliana Campos Junqueira

Abstract Oral candidiasis is an opportunistic infection caused by yeast of the *Candida* genus, primarily *Candida albicans*. It is generally associated with predisposing factors such as the use of immunosuppressive agents, antibiotics, prostheses, and xerostomia. The development of research in animal models is extremely important for understanding the nature of the fungal pathogenicity, host interactions, and treatment of oral mucosal *Candida* infections. Many oral candidiasis models in rats and mice have been developed with antibiotic administration, induction of xerostomia, treatment with immunosuppressive agents, or the use of germ-free animals, and all these models has both benefits and limitations. Over the past decade, invertebrate model hosts, including *Galleria mellonella, Caenorhabditis elegans*, and *Drosophila melanogaster*, have been used for the study of *Candida* pathogenesis. These invertebrate systems offer a number of advantages over mammalian vertebrate models, predominantly because they allow the study of strain collections without the ethical considerations associated with studies in mammals. Thus, the invertebrate models may be useful to understanding of pathogenicity of *Candida* isolates from the oral cavity, interactions of oral microorganisms, and study of new antifungal compounds for oral candidiasis.

Oral Candidiasis

Candida species is an innocuous commensal of the microbial communities of the human oral cavity. Its primary locations are the posterior tongue and other oral sites, such as the mucosa, whereas the film that covers the dental surfaces is colonized secondarily. Frequently, when the host defense system is compromised, *C. albicans* becomes virulent, resulting in the disease candidiasis, which may spread to multiple oral sites, or infect the entire oral cavity (Webb et al. 1998; Salerno et al. 2010). The presentations of oral candidiasis can be classified in pseudomembranous, erythematous, plaque-like (nodular) and *Candida*-associated lesions: angular cheilitis and median-rhomboid glossitis (Niimi et al. 2010).

Of the many pathogenic *Candida* species, *C. albicans* is the major fungal pathogen of humans. In the oral cavity, a niche they frequently inhabit as commensals, these yeasts exist predominantly

J.C. Junqueira (✉)
Department of Biosciences and Oral Diagnosis, School of Dentistry of São José dos Campos, UNESP-Univ Estadual Paulista, São Paulo, Brazil
email: juliana@fosjc.unesp.br

E. Mylonakis et al. (eds.), *Recent Advances on Model Hosts*,
Advances in Experimental Medicine and Biology 710, DOI 10.1007/978-1-4419-5638-5_10,

within biofilms: are spatially organized heterogeneous communities of fungal cells encased in a matrix of extracellular polymeric substances (Jin et al. 2004). The developing biofilms are constantly exposed and bathed in saliva. The effect of saliva on *Candida* adhesion and subsequent biofilm formation is controversial. Some studies have shown that saliva reduces the adhesion of *C. albicans* to oral surfaces. Indeed, saliva possesses defensive molecules, including lysozyme, lactoferrin, calprotectin and IgA, that decrease the adhesion of *Candida* to the oral surfaces. In other studies, salivary proteins such as the mucins and the statherins have been shown to act as adhesion receptors used by the mannoproteins present in the *Candida* species (Salerno et al. 2010).

The incidence of superficial and deep-seated fungal infections have increased markedly over the last 20 years. Several reasons have been proposed for the rise in incidence of fungal infections, including HIV infection, increasing use of immunosuppressive drugs, use of broad-spectrum antibiotics, prosthetic devices and grafts (Donelly et al. 2007). *Candida* strains have been isolated from 93% of patients with denture stomatitis, which is now considered to be the most common form of oral candidiasis (Noumi et al. 2010). Deng et al. (2010) verified that the incidence of oral candidiasis during radiation therapy in patients with head and neck cancer was significantly higher (55.2%) than in a nonirradiated control group (11.8%).

Oropharyngeal candidiasis (OPC) is the most frequent opportunistic infection encountered in HIV-infected individuals. OPC occurs in up to 90% of HIV-infected patients at some point during the course of HIV disease. The occurrence of OPC is associated with CD4 T-lymphocyte counts below 200 cells/mm^3, high viral loads, and disease progression (Hamza et al. 2008).

Several known virulence factors contribute to the pathogenicity of *C. albicans*, including the ability to adhere to epithelial cells, including the ability to form hyphae, and the secretion of extracellular enzymes (Noumi et al. 2010; Jayatilake et al. 2005; Pukkila-Worley et al. 2009). During the initial stages of superficial mucosal infection, *C. albicans* forms filamentous hyphae that show thigmotropism, a phenomenon also known as contact guidance, and release hydrolytic enzymes, such as extracellular phospholipases and secretory aspartyl proteinases (Jayatilake et al. 2005).

The widespread use of topical and systemic antifungal agents as the conventional treatment for oral candidiasis has resulted in the development of resistance in *C. albicans*. Although resistance of *C. albicans* to polyenes is rare, several mechanisms of azole resistance have been reported, including changes in the cell wall or plasma membrane, which decrease azole uptake, overexpression of or mutations in the target enzyme of azoles, and the efflux of azole drugs mediated by membrane transport proteins (Mima et al. 2010).

The versatility of the pathogenic mechanisms of fungi and their development of resistance to antifungal drugs indicate the importance of understanding the nature of these host-pathogen interactions in experimental systems, and warrant development of vertebrate and invertebrate models.

Mammalian Animal Models for the Study of Oral Candidiasis

General Aspects of Animal Models

Apart from the ethical dilemmas associated with experimentation on live humans, humans are notoriously different in terms of their dietary habits, social habits, immune status, and oral physiology, including salivary function. These factors, in addition to racial, ethnic, and cross-cultural variations in human demographics, influence the etiology and pathogenesis of diseases such as candidiasis. The development of an ideal animal model for oral candidiasis would provide a standardized tool which can be controlled and manipulated to derive universally comparable data on the etiopathology, diagnosis, and management of the disease process (Samaranayake and Samaranayake 2001).

Although many of the pioneering studies on mucosal candidiasis were performed in non-human primates, small mammals, including rats and mice, are the common choice for such studies for economical and ethical reasons and because of their relative anatomic and immunological similarity to humans (Samaranayake and Samaranayake 2001; Chamilos et al. 2007; Naglik et al. 2008).

Two species of rats, Sprague–Dawley and Wistar, have been widely used in experimental oral *Candida* infections. The main advantages of the rat model are the low maintenance cost and the sufficient size of the oral cavity, which easily permits *Candida* inoculation and sample collection. Furthermore, the tongue of this animal is fairly easily colonized by *Candida*, eliciting disease conditions such as median rhomboid glossitis and atrophic candidiasis (Samaranayake and Samaranayake 2001).

Mouse models are ideal for unraveling adaptive immune responses of the mucosal tissues to *Candida* infection because the immunobiology of the healthy murine oral mucosa has been fairly well characterized by a number of researchers. Furthermore, mice are easily obtained in large numbers and their maintenance is inexpensive (Samaranayake and Samaranayake 2001).

One caveat to studying *C. albicans*-host interactions in rodents is that *C. albicans* is not a natural colonizer of mucosal surfaces in these animals. The rodent equivalent of normal flora yeast is *Candida pintolopessi*, which can sometimes cause disease in immune-compromised rodents. This has both benefits and limitations. An advantage is that any host response to *C. albicans* is not affected by pre-existing adaptive immune responses to the fungus (Naglik et al. 2008). A disadvantage is that the establishment of mucosal colonization or infection usually requires intervention with antibiotics, induction of xerostomia, treatment with immunosuppressive agents, or the use of germ-free or transgenic animals.

Pros and Cons of Rat and Murine Models

Many oral candidiasis models in rats have been developed with antibiotic administration followed by persistent infections. The most commonly examined drug has been tetracycline because of its broad spectrum of activity and its association with human candidal infection (Allen 1994). A variety of protocols have been evaluated with varying parameters, including the dose of the antibiotic and the schedule of its administration. For induction of experimental oral candidiasis in several studies, the animals were treated daily with a solution of 0.08–0.1% tetracycline hydrochloride in their drinking water. This treatment was initiated 7 days prior to inoculation with *C. albicans* suspension and was maintained throughout the experiment (Allen et al. 1982; Fisker et al. 1982; Junqueira et al. 2009).

Oral candidiasis can also be induced by xerostomia, either by means of pharmacologic agents or the surgical removal of the salivary glands. One of the first studies of oral candidiasis in an animal model used hyoscine hydrobromide in an attempt to induce a xerostomic state in Wistar rats to increase the likelihood of infection (Jones and Adams 1970). In this instance, the use of the xerostomic drug did not seem to have much impact on the incidence of infection compared with untreated controls (Allen 1994). Further studies that used the ligation or removal of major salivary glands (parotid, sublingual, and submandibular) in the rat have shown an increased severity of infection in the xerostomic animals compared with normal controls (Jorge et al. 1993; Totti et al. 1996; Green et al. 2006). Jorge et al. (1993) verified that after 32 weeks of *C. albicans* inoculation, 20% of normal and 70% of sialoadenectomized rats showed candidal infection on the tongue.

Because the broad-spectrum antibiotics used, such as tetracycline, eradicate the antagonistic population pressure of the commensal oral flora, the sialoadenectomy model appears preferable for the investigation of oral candidiasis because it maintains the normal oral flora with its competitive, colonization pressure akin to the clinical conditions in humans (Samaranayake and Samaranayake 2001).

The antibiotic therapy and hyposalivatory models are limited by the lack of local symptoms characteristic of oral candidiasis, particularly oral thrush, in humans. Takakura et al. (2003)

developed a murine model of thrush-type oral candidiasis that mimics the natural infection in humans by immunosuppression with injections of prednisolone and tetracycline hydrochoride in drinking water. These authors induced typical candidiaisis lesions consisting of white patches on the tongue dorsum comparable to the pseudomembrane observed in human thrush, with extensive colonization on the epithelium by numerous hyphae and destruction of several epithelial layers.

Candidiasis-related lesions observed in rats models employing antibiotic therapy and hyposalivatory conditions are characterized by areas of papillary atrophy in the tongue dorsum without white patches, because the rats in these models have an intact immune system. The basis of the rat model is the profound change in ecology of the oral cavity, whereas immunosuppression is the basis for disease in the mouse model. The rat model represents what would be expected in individuals with chronically irritated mouths like Sjogren's syndrome patients, while the murine model reflects the oral ecology for an AIDS patient (Green et al. 2006).

Germ-free or gnotobiotic mice appear uniquely suited for studies of mucosal candidiasis, because mucosal surfaces can be naturally and chronically colonized by *C. albicans* without the need for sialoadenectomy, antimicrobial therapy, or immunosuppression. In addition, colonization can persist for the lifetime of the animal. However, the germ-free/gnotobiotic mouse is not ideal for the study of fungal dissemination from mucosal sites or natural host-pathogen interactions, because the absence of the normal microbiota is likely to have a major effect on both fungal pathogenicity and the host response (Naglik et al. 2008).

Transgenic mice are also extremely useful for experimental studies of sex-linked anemia, metabolic disease resembling diabetes mellitus in humans, and severe combined immune deficiency (SCID syndrome). The utilization of transgenic or congenitally immunodeficient mice has been instrumental in advancing our understanding of the critical roles of $CD4^+$(helper) T cells, $CD8^+$ (cytotoxic) T cells, polymorphonuclear leukocytes, macrophages, and cytokines in host defense against oral candidiasis (Naglik et al. 2008; He et al. 2010).

Microscopic Aspects of Experimental Oral Candidiasis

The microscopic features of experimental oral candidiasis in the tongue dorsum of rats with intact immune systems have been described by Junqueira et al. (2005). These authors observed yeasts and hyphae agglomerates in the keratin layer of epithelium by optical microscopy (MO) 6 h after inoculation with *Candida* yeasts. Many polymorphonuclear leucocytes in the prickle and basal cell layers were also observed. In scanning electron microscopy (SEM), only yeasts were found among the filiform papillae of the tongue dorsum. The hyphae could not be observed by SEM because they were present in the interior of the keratin.

After 24 h of *C. albicans* inoculation, the MO analysis showed high levels of hyphae and keratin desquamation. *Candida* infection increased the number of mitotic cells per unit length of the basal layer of the epithelium and the amount of desquamation. These changes were defense mechanisms of the rat against fungal invasion. The keratin desquamation allowed the observation of the large quantity of hyphae by SEM.

Seven days after *C. albicans* inoculation, the authors observed the presence of a small quantity of hyphae and of tissue lesions characterized by loss of filiform papillae, acanthosis and epithelium hyperplasia by MO. Although infection by *Candida* was restricted to the keratin layer in the epithelial surface, tissue changes occurred in the deepest layers of the epithelium. These effects can likely be attributed to defense mechanisms and extracellular enzyme production, such as proteinase and phospholipase. In the SEM examination, areas of atrophy and destruction of the filiform papillae and increased interpapillar surface area were observed.

The normal morphology of the tongue dorsum of rat and the development of experimental candidiasis can be observed in Figs. 10.1–10.6.

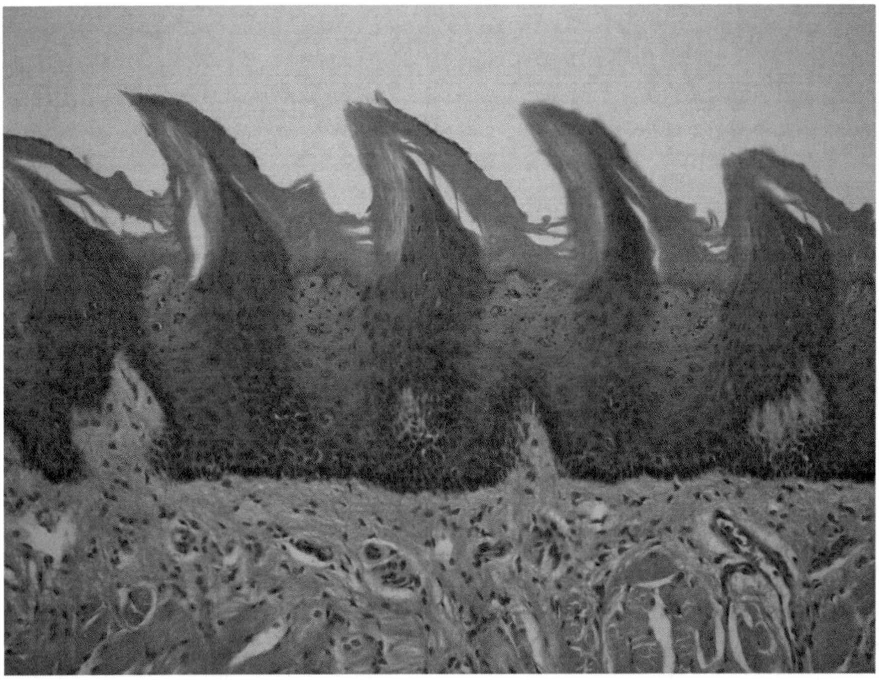

Fig. 10.1 Normal morphology of the tongue dorsum of rat. Optical microscopy: 200×

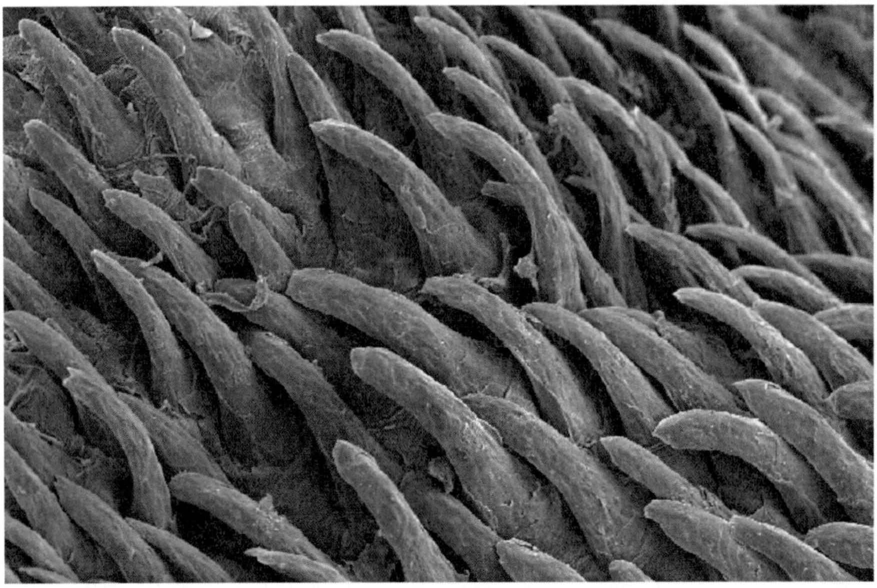

Fig. 10.2 Normal morphology of the tongue dorsum of rat. Scanning electron microscopy: 180×

Current Studies of Rat and Murine Models

Animal models of oral candidiasis have provided a wealth of information with respect to this disease process. Currently, they are used in the investigation of the pathogenicity, host interactions, and treatment of oral mucosal *Candida* infections.

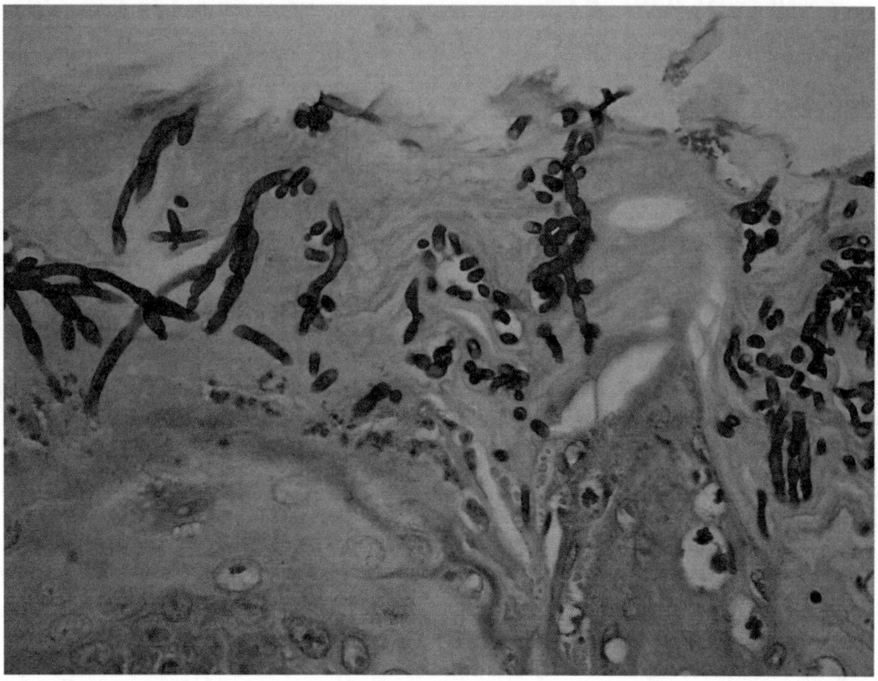

Fig. 10.3 Sagittal cut of the tongue dorsum of rat 6 h after C. albicans inoculation. Yeast, hyphae and polymorphonuclear leucocytes can be observed. Optical microscopy: 630×

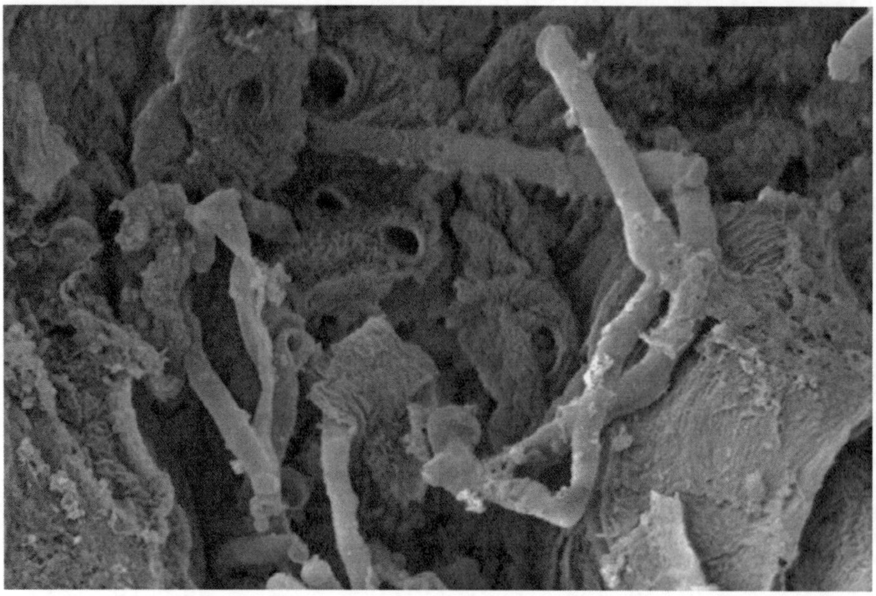

Fig. 10.4 Tongue dorsum of rat 24 h after C. albicans inoculation. Hyphae and tissue degradation are observed. Scanning electron microscopy: 2,500×

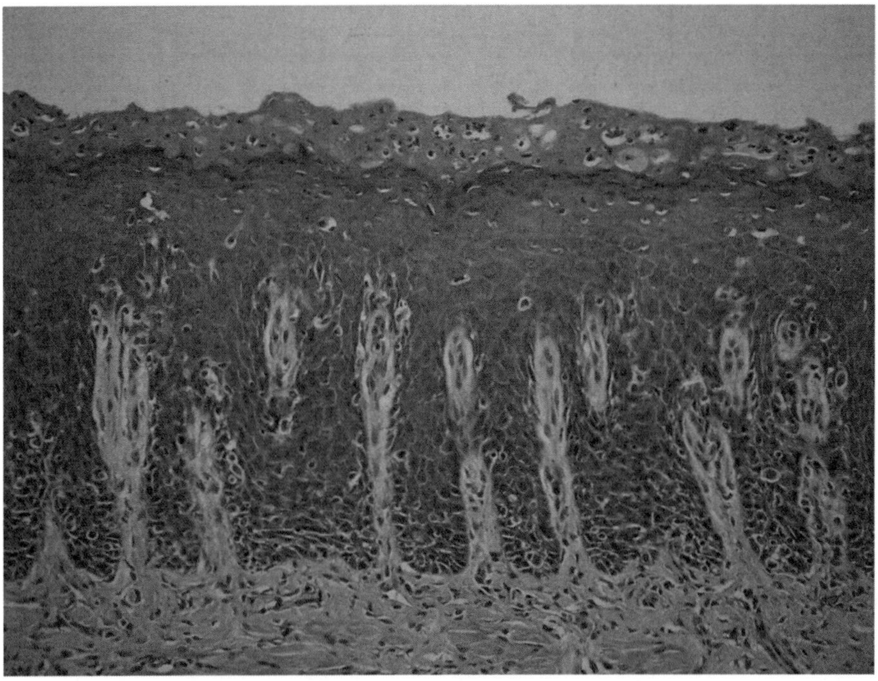

Fig. 10.5 Sagittal cut of the tongue dorsum of rat 7 days after C. albicans inoculation. Tissue lesion characterized by loss of filiform papillae, hyperparakeratosis and epithelium hyperplasia are observed. Optical microscopy: 200×

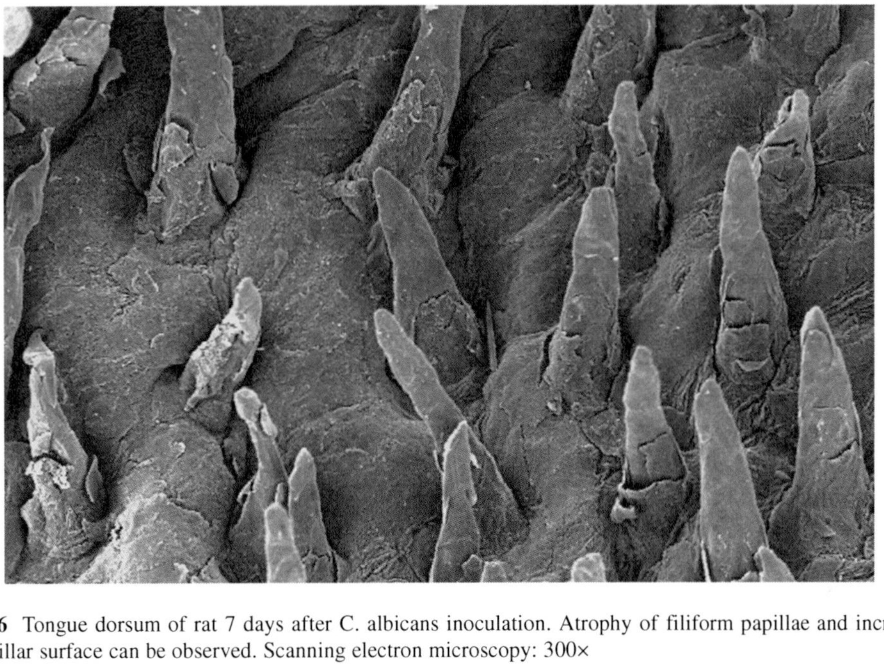

Fig. 10.6 Tongue dorsum of rat 7 days after C. albicans inoculation. Atrophy of filiform papillae and increase of interpapillar surface can be observed. Scanning electron microscopy: 300×

Green et al. (2006) studied *C. albicans* expression of ALS genes, encoding large cell-surface glycoproteins that function in the process of adhesion to host surfaces in the hyposalivatory rat model after surgical removal of the salivary glands. These authors verified that patterns of ALS gene expression were similar between the rat model and human clinical specimens, suggesting that the model could be used to study the phenotypes of *als/als* mutant strains.

He et al. (2010) reviewed several studies silencing or disrupting specific genes in knockout mice and concluded that the regulatory pathways governing *C. albicans* virulence are different during oropharyngeal candidiasis (OPC) and hematogenously disseminated candidiasis (HDC). For example, the *C. albicans* TPK2 (Takashi's protein kinase 2) gene product is one of two catalytic subunits in protein kinase A. The Ras-protein kinase A pathway regulates multiple processes in *C. albicans*, including hyphal formation, phenotypic switching, and clamydospore formation. Park et al. (2005) found that a tpk2D mutant of *C. albicans* had significantly reduced virulence in the mouse model of OPC, but normal virulence in the mouse model of HDC was maintained. According He et al. (2010) future studies are needed to determine whether there are differences in the specific virulence factors required for these distinct disease states.

Animal models have also been used to develop new therapies for oral candidiasis. Junqueira et al. (2009) and Mima et al. (2010) verified that photodynamic therapy promoted significant reduction of oral candidiasis in both the rat model with intact immune system and in an immunosuppressed mouse model. Taguchi et al. (2010) examined the therapeutic activity of spices and herbs for oral candidiasis in a murine model and suggested that oral intake of *Cinnamomum cassia* preparation is a clinical candidate for a therapeutic tool against oral *Candida* infection. The development of research with new therapeutic approaches in mammalian animal models is extremely important because the action of therapeutic agents on microorganisms could be affected by the environmental conditions of the oral cavity, including the presence of saliva, pH variations, mucosa characteristics and the action of the immunological system (Kömerik et al. 2003).

Invertebrate Models for the Study of Oral Candidiasis

Over the past decade, invertebrate mini-host models with well-characterized genetics and simple immunity have been effectively used to explore several aspects of both fungal pathogenicity and host immune response. Several factors sparked the development of these models. First, the mammalian animal models remain logistical barriers to large-scale studies. Second, the realization that innate immune mechanisms are evolutionarily conserved between invertebrates and mammals and that several common virulence factors are involved in fungal pathogenesis in phylogenetically disparate hosts – e.g., fruit flies, nematodes, and mammals – further expanded the field. Third, invertebrate organisms have been increasingly used as in-vivo assays for antifungal drug efficacy studies because of their low cost and simplicity (Chamilos et al. 2007).

Furthermore, because the genome sequences of medically important fungi such as *Aspergillus*, *Candida*, and *Cryptococcus* have been completed, the increase in genetic information has created an increasing need for simple innovative ways to screen for virulence mechanisms and assess the contribution of individual genes to fungal pathogenesis (Chamilos et al. 2007).

Model hosts, including the grater wax moth *Galleria mellonella*, the roundworm *Caenorhabditis elegans*, and the fruit fly *Drosophila melanogaster*, have been used for the study of *Candida* pathogenesis.

Galleria mellonella have been successfully used as models of *Candida* pathogenesis because of their relatively large size (about 5 cm long), which allows for the injection of standardized fungal inocula and studies of drug pharmacodynamics. In addition, *G. mellonella* can be maintained under various temperature conditions ranging from 25°C to 37°C. The ability to study pathogens at 37°C enables the study of temperature-related virulence traits. *Galleria* presents haemolymph with six types of hemocytes (prohemocytes, coagulocytes, spherulocytes, oenocytoids, plasmatocytes and granulocytes), which play a role in fungal-pathogen defense. The major disadvantage of these insects is the absence of methods for genetic analysis and the lack of a full genome sequence for this model (Chamilos et al. 2007; Fuchs and Mylonakis 2006; Mylonakis 2008).

Fuchs et al. (2010) evaluated the roles of five genes BCR1, FLO8, KEM1, SUV3 and TEC1, in *C. albicans* filamentation using a *G. mellonella* model. Among the five mutant strains tested, the authors observed that only the flo8/flo8 mutant strain did not form filaments within *G. mellonella*. This strain also exhibited reduced virulence in *G. mellonella* larvae. Another strain that exhibited reduced pathogenicity in the *G. mellonella* model was tec1/tec1 but, by contrast, the tec1/tec1 strain retained the ability to form filaments, suggesting that filamentation alone is not sufficient to kill *G. mellonella* and that other virulence factors may be associated with genes that regulate filamentation.

Rowan et al. (2009) employed larvae of *G. mellonella* to assess the in vivo antifungal efficacy of ($[Ag_2(mal)(phen)_3]$), $AgNO_3$ and 1,10-phenanthroline. Larvae pre-treated with these compounds were protected from a subsequent lethal infection by the yeast *C. albicans,* and larvae treated 1 and 4 h post-infection showed significantly increased survival compared to control larvae. Administration of these compounds resulted in an increase over 48 h in the density of insect hemocytes. These results demonstrate an increase in hemocyte numbers, which could contribute to the ability of the insect to kill *C. albicans;* this may function in combination with the antifungal properties of the compounds.

Caenorhabditis elegans is much smaller (about 1 mm long) than all other mini-host models. Nevertheless, *C. elegans* has a sequenced genome and fully developed genetic tools. Additionally, its hermaphroditic lifestyle and short lifespan (2–3 weeks) facilitate genetic experiments with this organism. *C. elegans* is maintained in Petri dishes and infected by ingesting the pathogen as a substitute for the usual laboratory food source, an auxotrophic strain of *Escherichia coli*. The simplicity of experimental infection in *C. elegans* allows for individual screening of thousands of virulence genes and candidate antimicrobial compounds (Pukkila-Worley et al. 2009; Chamilos et al. 2007, 2009; Fuchs and Mylonakis 2006).

Coleman et al. (2010) performed a compound screen to identify potential antifungal natural products using *C. elegans* model. Of the 12 antifungal saponins identified, two saponins (A7 and A20) were as effective as amphotericin B in promoting *C. elegans* survival. According to the authors, these compounds represent an opportunity to expand the current classes of antifungal agents in use and to improve available antifungal drugs by exploiting these new chemical scaffolds.

Drosophila melanogaster is larger (about 3 mm long) than *C. elegans* and is substantially smaller than *G. mellonella.* In the laboratory, *D. melanogaster* can be infected with fungal pathogens using various methods such as injection, direct spraying of fungal spores onto the flies, and ingestion. The genetic tractability and well-characterized immune system of Drosophila is a major advantage. The Drosophila genome sequence was one of the first to be completed and is one of the most fully annotated eukaryotic genomes available (Chamilos et al. 2007, 2009; Fuchs and Mylonakis 2006).

Chamilos et al. (2006) hypothesized that *C. albicans* has developed evolutionary conserved mechanisms to invade disparate hosts and tested whether *Toll* mutant flies of *D. melanogaster* could serve as a model host for high-throughput screening of *C. albicans* virulence genes. Of the 34 *C. albicans* mutants tested, only the prototrophic *cas*2 , mutant exhibited attenuated virulence in *To*ll mutant flies. Similarly, BALB/c mice infected intravenously with the *cas*2 , mutant had significantly better survival and a lower fungal burden in kidneys and spleen than did those infected with the isogenic wild-type strain. *CAS*5 encodes a key transcriptional regulator of genes involved in cell wall integrity. These findings support the notion that *D. melanogaster* is a promising model for large-scale studies of genes involved in the pathogenesis of *C. albicans* infection in mammals.

Future Directions and Conclusion

Experimental mammalian animal models for the study of oral candidiasis have been much explored, and they are important tools in assessing fungal pathogenicity, host immune defenses and treatment of oral mucosal *Candida* infections. However, there are no studies relating oral candidiasis and invertebrate models. Future research of oral candidiasis using invertebrate models are required to gain a better understanding of pathogenicity of *Candida* isolates from the oral cavity, interactions of oral microorganisms, and study of new antifungal compounds, since invertebrate models allow screening of virulence genes and antimicrobial compounds with low cost and simplicity.

References

Allen CM (1994) Animal models of oral candidiasis. A review. Oral Surg Oral Med Oral Pathol 78:216–221

Allen CM, Blozis GG, Rosen S, Bright RS (1982) Chronic candidiasis of the rat tongue: a possible model for human Median Rhomboid Glossitis. J Dent Res 61:287–291

Chamilos G et al (2006) *Drosophila melanogaster* as a facile model for large-scale studies of virulence mechanisms and antifungal drug efficacy in *Candida* species. J Infect Dis 193:1014–1022

Chamilos G, Lionakis MS, Lewis RE, Kontoyiannis DP (2007) Role of mini-host models in the study of medically important fungi. Lancet Infect Dis 7:42–55

Chamilos G, Nobile CJ, Bruno VM, Lewis RE, Mitchell AP, Kontoyiannis DP (2009) *Candida albicans* Cas5, a Regulator of cell wall integrity, is required for virulence in munrine and *Toll* mutant fly models. J Infect Dis 200:152–157

Coleman JJ et al (2010) Characterization of plant-derived saponin natural products against *Candida albicans*. ACS Chem Biol 5:321–332

Deng Z, Kiyuna A, Hasegawa M, Nakasome I, Hosokawa A, Suzuki M (2010) Oral candidiasis in patients receiving radiation therapy for head and neck cancer. Otolaryngol Head Neck Surg 143:242–247

Donelly RF, Mcarrin PA, Tunney MM, Woolfson AD (2007) Potential of photodynamic therapy in treatment of fungal infections of the mouth. Design and characterisation of a mucoadhesive patch containing toluidine blue O. J Photochem Photobiol B 86:59–69

Fisker AV, Rindon-Schiott C, Philipsen HP (1982) Long-term oral candidosis in rats. Acta Pathol Microbiol Immunol Scand 90:221–227

Fuchs BB, Mylonakis E (2006) Using non-mammalian hosts to study fungal virulence and host defense. Curr Opin Microbiol 9:346–351

Fuchs BB, Eby J, Nobile CJ, El Khoury JB, Mitchell AP, Mylonakis E (2010) Role of filamentation in Galleria mellonella killing by Candida albicans. Microbes Infect 12:488–496

Green CB, Marretta SM, Cheng G, Faddoul FF, Ehrharts EJ, Hoyer LL (2006) RT-PCR analysis of *Candida albicans* ALS gene expression in a hyposalivatory rat model of oral candidiasis and in HIV-positive human patients. Med Mycol 44:103–111

Hamza OJM et al (2008) Species distribuition and *in vitro* antifungal susceptibility of oral yeast isolates from Tanzanian HIV infected patients with primary and recurrent oropharyngeal candidiasis. BMC Microbiol 8:135

He H, Cong Y, Yang H, Dong Y (2010) Mutative expression in *Candida albicans* infection and cytokine signaling network in gene knockout mice. Eur J Clin Microbiol Infect Dis 9:913–916

Jayatilake JA, Samaranayake YH, Samaranayake LP (2005) An ultrastructural and a cytochemical study of candidal invasion of reconstituted human oral epithelium. J Oral Pathol Med 34:240–246

Jin Y, Samaranayake LP, Samaranayake Y, Yip HK (2004) Biofilm formation of *Candida albicans* is variably affected by saliva and dietary sugars. Arch Oral Biol 49:789–798

Jones JH, Adams D (1970) Experimentally induced acute oral candidosis in the rat. Br J Dermatol 83:670–673

Jorge AO, Totti MA, Almeida OP, Scully C (1993) Oral candidosis established in the sialoadenectomized rat. J Oral Pathol Med 22:54–56

Junqueira JC, Colombo CE, Martins JS, Koga-Ito CY, Carvalho YR, Jorge AOC (2005) Experimental candidosis and recovery of *Candida albicans* from the oral cavity of ovariectomized rats. Microbiol Immunol 49:199–207

Junqueira JC, Martins JS, Faria RL, Colombo CED, Jorge AOC (2009) Photodynamic therapy for the treatment of buccal candidiasis in rats. Lasers Med Sci 24:877–884

Kömerik N, Nakanishi H, MacRobert AJ, Henderson B, Speight P, Wilson M (2003) *In vivo* killing of *Porphyromonas gingivalis* by Toluidine Blue-mediated photosensitization in an animal model. Antimicrob Agents Chemother 47:932–940

Mima EG et al (2010) Susceptibility of *Candida albicans* to photodynamic therapy in a murine model of oral candidiasis. Oral Surg Oral Med Oral Pathol Oral Radiol Endod 109:392–401

Mylonakis E (2008) *Galleria mellonella* and the study of fungal pathogenesis: making the case for another genetically tractable model host. Mycopathologia 165:1–3

Naglik JR et al (2008) Animal models of mucosal *Candida* infection. FEMS Microbiol Lett 283:129–139

Niimi MN, Firth NA, Cannon RD (2010) Antifungal drug resistance of oral fungi. Odontology 98:15–25

Noumi E et al (2010) Adhesive properties and hydrolytic enzymes of oral *Candida albicans* strains. Mycopathologia 169:269–278

Park H et al (2005) Role of the fungal Ras-protein kinase A pathway in governing epithelial cell interactions during oropharyngeal candidiasis. Cell Microbiol 7:499–510

Pukkila-Worley R, Peleg AY, Tampakakis E, Mylonakis E (2009) *Candida albicans* hyphal formation and virulence assesses using a *Caernorhabditis elegans* infection model. Eukaryot Cell 8:1750–1758

Rowan R, Moran C, McCann M, Kavanagh K (2009) Use of *Galleria mellonella* larvae to evaluate the *in vivo* antifungal activity of $[Ag_2(mal)(phen)_3]$. Biometals 22:461–467

Salerno C et al (2010) *Candida*-associated denture stomatitis. Med Oral Patol Oral Cir Bucal 16:139–143

Samaranayake YU, Samaranayake LP (2001) Experimental oral candidiasis in animal models. Clin Microbiol Rev 14:398–429

Taguchi Y et al (2010) Therapeutic effects on murine oral candidiasis by oral administration of Cassia (*Cinnamomum cassia*) preparation. J Med Mycol 51:13–21

Takakura N et al (2003) A novel murine model of oral candidiasis with local symptoms characteristic of oral thrush. Microbiol Immunol 47:321–326

Totti MAG, Santos EB, Almeida OP, Scully C (1996) Implantation of *Candida albicans* and other *Candida* species in the oral cavity of rats. J Oral Pathol Med 25:308–310

Webb BC, Thomas CJ, Willcox MH, Harty DW, Knox KW (1998) *Candida*-associated denture stomatitis. Aetiology and management: a review. Aust Dent J 43:45–50

Chapter 11
Creating a Pro-survival and Anti-inflammatory Phenotype by Modulation of Acetylation in Models of Hemorrhagic and Septic Shock

Yongqing Li and Hasan B. Alam

Abstract Shock, regardless of etiology, is characterized by decreased tissue perfusion resulting in cell death, organ dysfunction, and poor survival. Current therapies largely focus on restoring tissue perfusion through resuscitation but have failed to address the specific cellular dysfunction caused by shock. Acetylation is rapidly emerging as a key mechanism that regulates the expression of numerous genes (epigenetic modulation through activation of nuclear histone proteins), as well as functions of multiple cytoplasmic proteins involved in key cellular functions such as cell survival, repair/healing, signaling, and proliferation. Cellular acetylation can be increased immediately through the administration of histone deacetylase inhibitors (HDACI).

A series of studies have been performed using: (1) cultured cells; (2) single-organ ischemia-reperfusion injury models; (3) rodent models of lethal septic and hemorrhagic shock; (4) swine models of lethal hemorrhagic shock and multi-organ trauma; and (5) tissues from severely injured trauma patients, to fully characterize the changes in acetylation that occur following lethal insults and in response to treatment with HDACI. These data demonstrate that: (1) shock causes a decrease in acetylation of nuclear and cytoplasmic proteins; (2) hypoacetylation can be rapidly reversed through the administration of HDACI; (3) normalization of acetylation prevents cell death, decreases inflammation, attenuates activation of pro-apoptotic pathways, and augments pro-survival pathways; (4) the effect of HDACI significantly improves survival in lethal models of septic shock, hemorrhagic shock, and complex poly-trauma without need for conventional fluid resuscitation or blood transfusion; and (5) improvement in survival is not due to better resuscitation but due to an enhanced ability of cells to tolerate lethal insults.

As different models of hemorrhagic or septic shock have specific strengths and limitations, this chapter will summarize our attempts to create "pro-survival and anti-inflammatory phenotype" in various models of hemorrhagic shock and septic shock.

Y. Li
Department of Surgery, Division of Trauma, Emergency Surgery and Surgical Critical Care, Massachusetts General Hospital/Harvard Medical School, Boston, MA, USA

H.B. Alam (✉)
Division of Trauma, Emergency Surgery, and Surgical Critical Care, Massachusetts General Hospital, 165 Cambridge Street, Suite 810, Boston, MA 02114, USA
e-mail: hbalam@partners.org

E. Mylonakis et al. (eds.), *Recent Advances on Model Hosts*,
Advances in Experimental Medicine and Biology 710, DOI 10.1007/978-1-4419-5638-5_11,

Introduction

Shock, such as hemorrhagic shock (HS) and septic shock, remains a major cause of morbidity and mortality among trauma patients and in intensive care units. Generally, HS-induced systemic response shares many features with septic response (Mollen et al. 2008). HS results in an early proinflammatory response, followed by a delayed generalized host immuno-suppression (Xu et al. 1998). Sepsis or septic shock is a complex syndrome that results from the host's inability to regulate the inflammatory response against infection. In hemorrhagic and septic shock, circulation is sub-optimal and host homeostasis is disturbed. At the molecular level, it has been reported that both hemorrhage and sepsis lead to an imbalance in protein acetylation and that HDACI can restore this balance (Lin et al. 2006; Gonzales et al. 2006; Li et al. 2009).

Lysine Acetylation and Histone Deacetylase Inhibitors

Lysine acetylation or N^{ε}-acetylation, identified initially on core histones in 1968 (Vidali et al. 1968), is mediated by a group of enzymes called histone acetyltransferases (HATs), which transfer acetyl groups from acetyl-coenzyme A to the ε-amino group of lysines. HATs are counterbalanced by the activity of histone deacetylases (HDACs) that catalyze the hydrolytic removal of the acetyl group of lysines. In humans, HDACs are divided into four classes (Table 11.1) based on their homology to yeast HDACs (Marks and Dokmanovic 2005; Carey and La Thangue 2006). Class I HDACs include HDAC1, 2, 3 and 8; these are related to the yeast enzyme Rpd3. Class II HDACs include HDAC4, 5, 6, 7, 9 and 10, and are related to the yeast protein HDA1 (histone deacetylase-A1). Class II HDACs are further divided into two subclasses – IIa (HDAC4, 5, 7 and 9) and IIb (HDAC6 and 10) – according to their structural similarities. Class III HDACs are referred to as sirtuins owing to their homology to the yeast HDAC Sir2. This class includes SIRT1–SIRT7 (Chuang et al. 2009; Suzuki 2009). HDAC11, the most recently identified isoform, is a class IV HDAC due to its distinct structure (Voelter-Mahlknecht et al. 2005). Class I, II, and IV are zinc-dependent enzymes, whereas class III HDACs are nicotinamide adenine dinucleotide (NAD+)-dependent enzymes. Based on their various sub-cellular localizations, intra-tissue variation and non-redundant activities, the different HDACs are implicated in various specific cellular processes, such as proliferation, metabolism and differentiation. For example, class I HDACs are mainly nuclear enzymes, whereas class II HDACs localize either to the cell nucleus or to the cytoplasm, depending on their phosphorylation and subsequent binding of 14-3-3 proteins. Moreover, class I HDACs are ubiquitously expressed (de Ruijter et al. 2003; Hu et al. 2003; Dangond and Gullans 1998; Mai et al. 2003), whereas class II HDACs display a tissue-specific expression.

To date, more than 15 HDACI have been tested in preclinical and early clinical studies for cancer therapy (Lane and Chabner 2009). Many of them are broad-spectrum or pan-HDACI, which inhibit many of the Class I, II and IV isoforms including suberoylanilide hydroxamic acid (SAHA), trichostatin A (TSA) and valproic acid (VPA). Some clinical compounds such as MS-275, FK-228 and apicidin have been termed as "Class I-selective", since they target several Class I isoforms of HDAC. Tubacin is one of a few HDACI that have been reported as a HDAC6 specific inhibitor (Haggarty et al. 2003) (Table 11.1). Thus far, nearly all of studies are focused on class I and class II HDACs, while very few are focused on class III/sirtuins (Weichert 2009).

Extracellular Signals and Protein Acetylation Balance

Signals that enter the cell nucleus encounter chromatin, and this interaction has a major impact on gene expression. Regulation of gene expression has two components that act in concert: alteration of chromatin structure governed by histone modification and binding by transcription factors (TFs)

Table 11.1 Classification of HDACs and selected HDACI

HDAC class	HDAC isoforms	Localization of HDAC	Specific HDAC inhibitors	Non-specific HDAC inhibitors	References
Class I	HDAC1	Nucleus	MS-275, FK-228	TSA	Chuang et al. (2009), Lane and Chabner (2009)
(Zn^{++} – dependent)	HDAC2	Nucleus	FK-228, apicidin	SAHA	
	HDAC3	Nucleus	MS-275, apicidin	Butyrate	
	HDAC8	Nucleus		Valproic acid	
Class IIa	HDAC4	Nuc/Cyt		TSA	Chuang et al. (2009), Lane and Chabner (2009)
(Zn^{++}-dependent)	HDAC5	Nuc/Cyt		SAHA	
	HDAC7	Nuc/Cyt		Butyrate	
	HDAC9	Nuc/Cyt		Valproic acid	
Class IIb	HDAC6	Mainly Cyt	Tubacin	TSA	Lane and Chabner (2009)
(Zn^{++}-dependent)	HDAC10	Mainly Cyt		SAHA	
Class III	SIRT1	Nuc/Cyt	Suramin	Nicotinamide	Chuang et al. (2009), Suzuki (2009), Scher et al. (2007)
(NAD^{+}-dependent)	SIRT2	Nuc/Cyt	Suramin, AGK2		
	SIRT3	Nuc/Mitoch			
	SIRT4	Mitochondria			
	SIRT5	Mitochondria			
	SIRT6	Nucleus			
	SIRT7	Nucleus			
Class IV (Zn^{++}-dependent)	HDAC11	Nuc/Cyt		TSA SAHA	Lane and Chabner (2009)

including activators and repressors. Most importantly, two classes of enzymes control the alteration and binding: HAT and HDAC. HAT modify core histone tails by post-translational acetylation of specific lysine residues and create an appropriate 'histone code' for chromatin modification to enhance DNA accessibility of TFs. In general, acetylation of core histones unpacks the condensed chromatin and renders the target DNA accessible to transcriptional machinery, hence contributing to gene expression. In most cases, the TFs can also be acetylated by HAT to facilitate their interactions with DNA and other proteins for transactivation. By contrast, deacetylation of the histones and TFs by HDAC increases chromatin condensation and precludes binding between DNA and TFs leading to transcriptional silencing.

Moreover, HAT and HDAC also target non-histone proteins, which may represent general regulatory mechanisms in biological signaling. In normal conditions, protein concentration and enzyme activity of HAT and HDAC remain in a highly harmonized state of balance. In an attempt to emphasize the importance of regulated acetylation this process is often referred to as 'acetylation homeostasis' (Saha and Pahan 2006). For example, recent studies have shown that various neurodegenerative challenges disturb this balance by decreasing HAT activity and the ratio tilts in favor of HDAC. The impaired acetylation homeostasis causes transcriptional dysfunction and facilitates a neurodegenerative cascade, which has been implicated in pathogenesis of several neurodegenerative disorders (Sugars and Rubinsztein 2003; Cha 2000). Indeed, perturbation of acetylation homeostasis is being recognized as a central event in the pathogenesis of neurodege-neration and HDAC inhibitors (HDACI) to be protective in animal models of Huntington's disease (Steffan et al. 2001; McCampbell et al. 2001; Ferrante et al. 2003; Hockly et al. 2003), amyotrophic lateral sclerosis (Ryu et al. 2005; Petri et al. 2006), experimental autoimmune encephalitis (Camelo et al. 2005), spinal muscular dystrophy (Chang et al. 2001; Avila et al. 2007) and other disorders (Yildirim et al. 2008).

Recently, HDACI have also been identified to be potent pro-survival and anti-inflammatory drugs, offering new lines of therapeutic intervention for hemorrhagic shock and septic shock. Our team and other groups have found that shock causes a global cellular hypoacetylation. Also HDACI, such as VPA, SAHA and TSA prevent hemorrhage-associated lethality in rat and swine models of hemorrhagic shock (Alam et al. 2009; Lin et al. 2006, 2007; Gonzales et al. 2006), suppress expression of proinflammatory cytokines and improve survival in a mouse model of septic shock (Li et al. 2009; Cao et al. 2008; Zhang et al. 2009). We have also demonstrated that inhibition of HDAC can modulate the immune response (trauma/hemorrhage and inflammatory second hit in-vitro) not only in animals, but also in severely injured patients (Sailhamer et al. 2008).

Models for Study of Hemorrhagic Shock and Septic Shock

So far, a broad variety of experimental conditions, including in vivo animal and in vitro cell-based models, have been established to enable investigators to study the effects of the hypovolemic shock. These models have provided researchers with different conditions to assess the potential benefits of a wide spectrum of treatment options.

In essence, there are two types of in vivo experimental models of hemorrhagic shock: controlled hemorrhage and uncontrolled hemorrhage. Uncontrolled-hemorrhage models are clinically realistic as the animals are allowed to bleed regardless of blood pressure or the volume of blood loss (similar to actual trauma patients). This clinical realism, however, introduces a huge degree of animal-to-animal variability. These models are used mostly for translational studies that focus on clinical outcome/ survival, and not when the investigators are interested in studying specific mechanisms or pathways. Controlled hemorrhage models are either of fixed-pressure or fixed-volume type (Lomas-Niera et al. 2005). In the former model, animals are bled to a predetermined mean arterial pressure (MAP) and are maintained at that pressure, with periodic bleeding, for a specified period of time based on the degree or outcome of hypotensive shock. In the later model, a controlled volume of blood (as a percentage of total circulating blood volume) is withdrawn. Other important variables must also be controlled in these models such the actual bleeding time, duration of shock, body size, gender, type of anesthesia, and presence or absence of resuscitation, just to name a few. These controlled models, although clinically unrealistic, are suitable for mechanistic studies as they allow the researchers to control most of the confounding variables. They are thus routinely used in the investigation of end organ damage, cardiovascular alterations, subsequent central nervous system and spinal cord injuries, immunological changes and response to fluid resuscitation (Moochhala et al. 2009).

As for the in vivo sepsis models, on the basis of the initiating agent, they can be divided into five broad categories: (1) exogenous administration of a toxin such as lipopolysaccharide (LPS), (2) intravenous infusion of a viable pathogen, (3) administration of fecal material or live organisms into the peritoneal cavity, (4) placement of infected foreign material into the soft tissues of the extremity, and (5) surgical operations that partially destroy the normal barrier of the gastrointestinal tract [e.g., cecal ligation and puncture (CLP) and colon ascendant stent peritonitis (CASP)] (Bhatia et al. 2009). Although many animal models of sepsis or septic shock have been developed, none replicate all the aspects of clinical sepsis. The advantages and disadvantages of these animal models are summarized in Table 11.2.

Cell and organ-based experiments are now being used as in vitro models and are offering direct molecular and cellular accessibility and micro-environmental controls. Additionally, they provide efficient comparison between many experimental conditions or potential therapeutic compounds. So far, in vitro models such as cultured cells in hypoxic condition or conditional medium (Li et al. 2009) and isolated organ (e.g., heart) perfused with different fluids (Zhao et al. 2007) have been established to analyze mechanisms of HDACI action involved in HS and septic shock.

As the different models of hemorrhagic and septic shock have specific advantages and disadvantages (Table 11.2), it is important to analyze the experimental findings in the context of the selected model. For this manuscript, we will focus on the fixed volume hemorrhage and LPS- induced septic shock (and some in vitro models), to highlight the emerging role of HDACI in the treatment of lethal insults.

HDACI in Models of Hemorrhagic Shock

HS creats a global ischemic insult due to acute blood loss. Current treatment for HS focuses on pathophysiology at the level of organ systems: maintain sufficient tissue perfusion and vital organ function through administration of fluids and blood products, and to surgically control the source of hemorrhage. Unfortunately, this resource intensive protocol remains difficult to administer, particularly in austere environments where advanced surgical interventions are not available (Champion et al. 2003). Moreover, this approach fails to address much of the damage that takes place at the cellular level as a result of hypoperfusion (during hemorrhage) and reperfusion (during resuscitation) (Valko et al. 2007).

In Vivo Rodent and Swine Models of Hemorrhagic Shock

Hemorrhage is responsible for about half of the trauma deaths, and most of these occur during the pre-hospital period. To alter this grim outcome, the first responders must keep the injured alive long enough to be transported to a hospital for definitive care. As the pre-hospital environment is fairly austere, especially in the battlefield, conventional resuscitation strategies that rely on transfusion of

Table 11.2 Animal models used for study of hemorrhagic shock and septic shock (Lomas-Niera et al. 2005; Bhatia et al. 2009)

	Animal models	Advantage	Disadvantage
Hemorrhagic shock	Fixed pressure hemorrhage	The extent of hypotension, duration and/or the volume can be controlled/monitored	Animal typically heparinized
	Fixed volume hemorrhage	Models acute hemorrhage (hypotension)	Degree of hypotension uncertain
	Uncontrolled hemorrhage	Considered most clinically relevant	No standardized control of degree/duration of hypotension and extent of blood loss
Septic shock	LPS-induced sepsis	1. Endotoxicosis model 2. LPS is convenient to use 3. Doses of LPS are readily measured	1. Single toxin may not mimic clinical sepsis 2. Variable hemodynamic responses with different doses and infusion rates 3. Administration route, LPS doses and host species may affect responses
	Bacterial infection (intravenous infusion of a viable bacteria, administration of fecal material or live organisms into peritoneal cavity)	Intravenous infusion of live organisms may be appropriate to study the blood clearance kinetics of organisms	1. Requires growth and quantification of bacteria prior to administration 2. Significant inter-laboratory variability 3. Host response id dependent on infecting bacterial strain and route of administration 4. Different host response with different compartment infection 5. Variable host response depends on bacterial load and infusion time 6. Large quantity of bacteria used may elicit confounding toxicosis response
	CLP-induced Sepsis	1. Simple and easy surgical procedure 2. Relatively reproducible 3. No need to prepare bacteria 4. Resemblance to clinical conditions	1. Multiple bacterial flora 2. Inter-laboratory variability 3. Sex and age variability 4. Strain variability 5. Needle size, number of punctures and amount of cecum ligated may affect the severity of sepsis 6. Difficult to control bacterial load and magnitude of sepsis challenge 7. Potential of ascribing sepsis therapy success to enhanced abscess formation mechanism 8. Human therapy potentially withheld could detract from validity of therapeutic agent
	CASP-induced Sepsis	Induce diffuse peritonitis with persistent systemic infection	1. More challenging surgical procedure and requirement for careful surgical techniques 2. Multiple bacteria flora 3. Less characterized hemodynamic response 4. Less experience to identify possible confounding variables 5. Mortality rate varies with stents of different diameters 6. Load of stool transferred into peritoneum may be a confounding variable of CASP 7. Human therapy potentially withheld could detract from validity of therapeutic agent

blood products are unrealistic. The challenge is for the researchers to develop novel strategies that can maintain life without fluid/blood resuscitation.

Shults and colleagues subjected male Wistar-Kyoto rats to 60% blood volume loss and treated them with or without HDACI such as VPA or SAHA (Shults et al. 2008). They then examined the rats over the next 3 h, using the survival as primary endpoint. The results showed this model was highly lethal as only 25% of the animals survived for 3 h. However, administration of HDACI after hemorrhage (no fluid or blood administration) significantly improved survival (75% and 83% in VPA and SAHA groups, respectively). These results demonstrated that post-shock treatment with HDACI can significantly improve early survival in a highly lethal model of hemorrhagic shock, even in the absence of conventional resuscitation. Following up on this subject, Alam et al. investigated whether VPA treatment would improve survival in a clinical relevant large animal (swine) model of poly-trauma/hemorrhagic shock (femur fracture, 60% blood loss, liver injury, hypothermia, acidosis and coagulopathy). They found that treatment with VPA without blood transfusion improved early survival in this highly lethal model (Alam et al. 2009). How can HDACI maintain organ viability without restoring intra-vascular volume and tissue perfusion? Although still incomplete, recent studies have markedly advanced our appreciation of the underlying mechanisms.

In Vitro Single-Organ Ischemia-Reperfusion Injury Model

Ischemia/reperfusion injury may lead to myocardial infarction, cardiac arrhythmias, and contractile dysfunction. The phenomenon of ischemic preconditioning, in which a period of sublethal injury can protect the cell during a subsequent ischemic insult, has been widely investigated over the last two decades. The preconditioning consists of an early and a late phase. The early phase, termed "early" or "first window," develops immediately and disappears within 1–2 h of ischemic preconditioning stimulus. Conversely, the late phase also known as the "second window" or "delayed," manifests after 12–24 h and lasts for 3–4 days (Ping and Murphy 2000). It is known that in the preconditioning stimuli a series of signal transduction pathways carry the signal for protection, and these presumably terminate on one or more end-effectors. In activity the end-effectors cause the protection during the lethal ischemic insult (index ischemia) and/or the subsequent reperfusion period. A memory element exists, somewhere in the signal transduction pathways between the trigger signal and the end-effector, that is set by the preconditioning protocol and keeps the heart in a preconditioned state (Yellon and Downey 2003).

To assess whether HDACI trigger preconditioning-like effects against ischemia/reperfusion injury, Zhao et al. isolated mouse hearts and perfused them with three cycles of 5-min infusion and 5-min washout of 50 nM of TSA, a potent inhibitor of HDAC, to mimic early pharmacologic preconditioning. This was followed by 30 min of ischemia and 30 min of reperfusion. In addition, mice were treated with saline or TSA (0.1 mg/kg, i.p.) to investigate delayed pharmacologic preconditioning. This study found that infusion of TSA itself did not change ventricular function in non-ischemic hearts. However, when the perfused heart was subjected to ischemia/reperfusion in vitro, TSA treatment resulted in an improvement in the recovery of ventricular function and a reduction in infarct size (Zhao et al. 2007).

The survival advantage is not due to improvement in resuscitation but to better tolerance of shock by the cells. The cell protective mechanisms may result from (1) epigenetic regulation through post translation modification of histone proteins, (2) activation of cell survival factors such as the phosphoinositide 3-kinase (PI3-k)/Akt signaling pathway, (3) blockage of gut-liver/lymph-lung axis, and/or (4) breakage of a paracrine loop between leukocytes and endothelial cells. All of these actions directly or indirectly involve restoration of protein acetylation.

Acetylation-Related Epigenetic Regulation

Modulation of histone acetylation to restore and maintain the normal ratio of HAT/HDAC (epigenetic regulation) has been tested as potential treatment for many diseases. A number of preclinical studies have demonstrated that HDACI can improve survival in degenerative diseases, prevent the brain from various insults, attenuate the effects of aging and increase life span (Ryu et al. 2003; Steffan et al. 2001; Chang and Min 2002; Faraco et al. 2006). Our group has reported that administration of HDACI protects organs and cells from hemorrhagic shock-induced injury (Shults et al. 2008; Li et al. 2008a; Sailhamer et al. 2008). Several converging lines of inquiry suggest that HDACI protect key organs by minimizing the cellular damage during hemorrhagic shock and resuscitation.

In the heart, ischemia induces deacetylation of histones H3/4 in vitro and in vivo (Granger et al. 2008). Using standard murine model of heart ischemia-reperfusion, Granger et al. demonstrated that treatment with HDACI significantly reduces infarct area, even when delivered 1 h after the ischemic insult. HDACI decrease the response to ischemic injury and lessen the size of myocardial infarction (Granger et al. 2008). In part, this is through prevention of ischemia-induced activation of gene programs that include hypoxia inducible factor-1α, cell death, and by decreasing vascular permeability in vivo and in vitro, which reduces vascular leak and myocardial injury.

In the liver, oxygen deprivation increases HDAC1, 4, and −5 protein levels by twofold and decreases acetylated histone H3 levels to 50–75% of the control values in a turtle model of anoxia (Kochenek et al. 2010). In a rat model of hemorrhagic shock, Gonzales et al. reported that hemorrhage increased serum levels of lactate, lactate dehydrogenase, aspartate aminotransferase, and alanine aminotransferase. Alternatively, treatment with VPA (an HDACI) induced acetylation of histones (H2A, H3, and H4), normalized serum levels of these enzymes and prolonged survival by fivefold. Furthermore, hyperacetylation of the histone proteins indicated the presence of active genes and correlated with improved survival (Gonzales et al. 2008). Gene expression profiling data from our group has shown that VPA treatment up-regulates expression of 17 critical genes at the early stage of HS (Fukudome et al. 2011). Two of these genes are peroxisome proliferator-activated receptor γ coactivator-1α (PGC-1α) and dual specificity protein phosphatase 5 (DUSP5). PGC-1α protects cells from oxidative stress by increasing the expression of various antioxidant defense enzymes including superoxide dismutase and glutathione peroxidase (St-pierre et al. 2006). DUSP5 is an inducible, nuclear, dual-specificity phosphatase, which specifically interacts with and inactivates the extracellular signal-regulated kinase (ERK) 1/2 MAP kinases in mammalian cells (Kucharska et al. 2009). Inactivation of ERK1/2 MAP kinases by DUSP5 could be one of mechanisms responsible for the protective properties of VPA in HS.

In the kidney, it has been discovered that ischemia/reperfusion induces a transient decrease in histone acetylation in proximal tubular cells. This is likely a result of a decrease in histone acetyltransferase activity as suggested by experiments with energy-depleted renal epithelial cells in culture (Marumo et al. 2008). During recovery after transient energy depletion in epithelial cells, the HDAC isozyme HDAC5 is selectively downregulated in parallel with the return of acetylated histone. Knockdown of HDAC5 by RNAi significantly increased histone acetylation and bone morphogenetic protein-7 (BMP7) expression (Marumo et al. 2008). In a rat model of HS, it was found that treatment of animal with VPA or SAHA markedly increases the phosphorylation of Akt and decreases the expression of pro apoptotic BAD (Bcl-xl/Bcl-2 associated death promoter) protein in kidney tissue (Zacharias et al. 2010). Further investigation is needed to find if there is any relationship between HDAC5 inhibition and Akt activation.

In the brain, Faraco et al. found that ischemia (6 h of middle cerebral artery occlusion) drastically decreases histone H3 acetylation levels without evidence of a concomitant change in histone acetyltransferase or deacetylase activities. Treatment with SAHA (50 mg/kg i.p.) increased histone H3 acetylation in the normal brain (approximately eightfold after 6 h) and prevented histone deacetylation in the ischemic brain. These effects were accompanied by increased expression of the neuroprotective Heat-shock protein 70 (Hsp70) and B-cell lymphoma 2 (Bcl-2) protein in the control and ischemic brain 24 h after the insult. At the same time point, mice injected with SAHA at 25 and 50 mg/kg had smaller infarct volumes compared with vehicle-treated animals (28.5% and 29.8% reduction, $p < 0.05$ versus vehicle). Recently, Li et al. reported that VPA treatment induces acetylation of histone H3, increases expression of β-catenin and Bcl-2 proteins, and prevents neuronal apoptosis in in-vitro hypoxic condition (0.5% O_2), as well as in in-vivo model of HS (Li et al. 2008a). These findings demonstrate that pharmacological inhibition of HDAC promotes expression of neuroprotective proteins within the ischemic brain, which underscores the therapeutic potential of these drugs.

Activation of Phosphoinositide 3-Kinase (PI3K)-Akt/PKB Pathway

Activation of PI3K enhances cell survival (and decreases apoptosis) via Akt/PKB activity in many cell types including cardiomyocytes, cardiac fibroblast, vascular smooth muscle cells (VSMCs), endothelial cells and hepatocytes (Oudit et al. 2004; Shuja et al. 2009; Alam et al. 2009). Alam et al. recently demonstrated that VPA exerts anti-apoptotic effect through the Akt/PKB signaling pathway to improve survival in a swine model of poly-trauma and massive blood loss (Alam et al. 2009). In this study they compared cell protective effects of the HDAC inhibitor (VPA) to treatment with fresh whole blood (FWB) as well as conventional normal saline resuscitation. VPA treatment increased the levels of activated Akt, deactivated glycogen synthase kinase-3β (GSK-3β), β-catenin and Bcl-2 significantly when compared to FWB and saline control groups (without an alteration in the total Akt and GSK-3β levels) (Alam et al. 2009). VPA has been reported to directly and indirectly inhibit GSK-3β (Kim et al. 2005). However, it is not clear how the HDACI influences the Akt signaling in an animal model of trauma and hemorrhagic shock. There are several other possibilities for the activation of this pathway in addition to the direct inhibition of GSK-3β (Fig. 11.1).

1. *An increase in the acetylated tribbles (TRB) 3 and phosphatase and tensin homolog (PTEN) may be associated with PI3K/Akt activity* (Yao and Nyomba 2008). Growth factors such as insulin, insulin-like growth factor 1 (IGF-1), erythropoietin, and cytokines that reduce apoptosis rely almost exclusively on the PI3K/Akt pathway, whereas GPCR-induced PI3K/Akt activation and cardioprotection occurs in response to several peptide agonists including urocortin, ghrelin, and adrenomedullin as well as beta2-adrenergic receptor (β_2-AR) stimulation (Oudit et al. 2004; Torella et al. 2004; Chesley et al. 2000; Kim et al. 2008). In the heart, overexpression of Akt/PKB causes resistance to apoptosis (Oudit et al. 2004); knockout of Akt2/PKBβ enhances apoptosis in response to myocardial ischemia (DeBosch et al. 2006). Consistent with a critical role for Akt/PKB in cell survival, loss or gain of TRB3 and PTEN activity leads to reduced or enhanced apoptosis, respectively (Avery et al. 2010; Kishimoto et al. 2003). Alternatively. increased expression of TRB3 and PTEN promotes apoptosis in cardiac myocytes (Schwartzbauer and Robbins 2001; Avery et al. 2010).

   PTEN is a dual protein/lipid phosphatase whose main substrate is phosphatidyl-inositol,3,4,5 triphosphate (PIP3), the product of PI3K. PTEN degrades PIP3 to an inactive form phosphatidylinositol 4,5-bisphosphate (PIP2) (Lee et al. 1999; Maehama and Dixon 1998; Oudit et al. 2004; Stambolic et al. 1998), inhibiting Akt activation. PTEN is constitutively active and is the major downregulator of PI3K/Akt (Stiles et al. 2004). PTEN also forms signaling complexes with PDZ domain-containing adaptors, such as the MAGUK (membrane-associated guanylate kinase) proteins. These interactions appear to be necessary for the metabolism of localized pools of PIP3 involved in regulating actin cytoskeleton dynamics. Acetylation is major mechanism that regulates PTEN activity (Okumura et al. 2006). Histone acetylase p300/CREB-binding protein-associated factor (PCAF) interacts with PTEN and acetylates lysines 125 and 128 which are

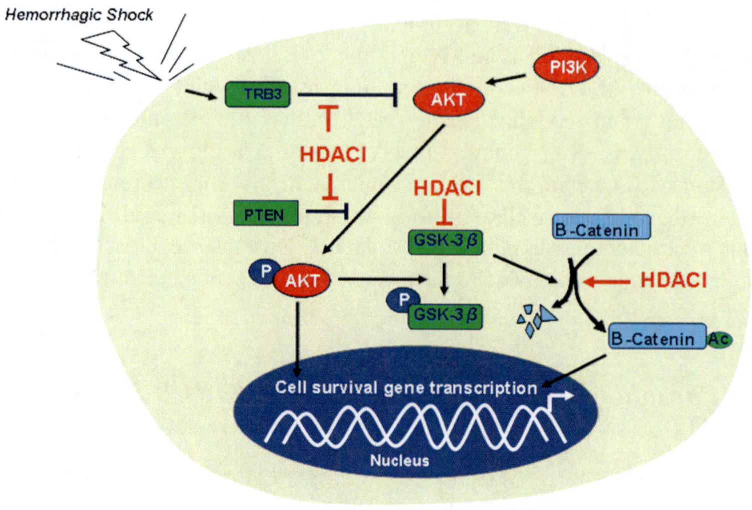

Fig. 11.1 Effect of HDACI on cell survival signaling pathway in hemorrhagic shock. HDACI induce phosphorylation of AKT by inhibition of TRB3 and PTEN. While AKT stimulates transcription of cell survival genes through several other pathways, phosphorylated AKT phosphorylates GSK-3β. Phosphorylated GSK-3β becomes inactivated form, which cannot degrade β-catenin. HDACI can also directly inhibit GSK-3β. Moreover, HDACI induce acetylation and nuclear translocation of β-catenin, leading to downstream survival gene transcription. *P* phosphorylation, *Ac* acetylation

located within the catalytic cleft of PTEN and are essential for PIP3 specificity. PCAF functions as a negative regulator of PTEN (Tamguney and Strokoe 2007). TRB3 is an intracellular pseudokinase that modulates the activity of several signal transduction cascades. TRB3 has been reported to inhibit the activity of Akt protein kinases (Du et al. 2003). TRB3 gene expression is highly regulated in many cell types, and hypoxia or endoplasmic reticulum (ER) stress promotes TRB3 expression. TRB3 binds to inactive and unphosphorylated Akt, thus preventing its phosphorylation (Shih et al. 2003). It remains unknown whether and how PCAF regulates TRB3. Recently Yao and Nyomba reported that acetylation status of TRB3 and PTEN is decreased in association with increased HDAC and decreased HAT activities. The hypoacetylated TRB3 and PTEN can inhibit the Akt-activity in a rat model of prenatal alcohol exposure (Yao and Nyomba 2008), which suggests that HDACI treatment could inhibit the activity of TRB3 and PTEN, which in turn would enhance the Akt signaling.

2. *Induction of Hsp70 by HDACI may be associated with PI3K/Akt pathway regulation.* In a rat model of hemorrhagic shock, Gonzales et al. found that VPA treatment increased the acetylation of nonhistone and histone proteins and expression of Hsp70 in rat myocardium, and significantly prolonged survival (fivefold) compared to the untreated controls (Gonzales et al. 2006). In rat cortical neurons, VPA treatment markedly up-regulated Hsp70 protein levels, and this was accompanied by increased Hsp70 mRNA levels and promoter hyperacetylation and activity (Marinova et al. 2009). Other HDAC inhibitors – sodium butyrate, trichostatin A, and Class I HDAC-specific inhibitors MS-275 and apicidin – all possess the ability to induce HSP70.

Hsp70 is a molecular chaperone, cell-protective and anti-inflammatory agent. Marinova et al. recently reported that HDACI increase Sp1 acetylation, promote the association of Sp1 with the histone acetyltransferases p300 and recruitment of p300 to the Hsp70 promoter. Further, HDACI-induced cell protection can be prevented by blocking Hsp70 induction (Marinova et al. 2009). In addition, Gao and Newton showed that Hsp70 directly binds and stabilizes Akt/PKB as well as protein kinase A and protein kinase C, thus prolonging the signaling lifetime of the kinases (Gao and Newton 2002). Taken together, these findings suggest that the PI3K/Akt pathway and Sp1 are likely involved in Hsp70 induction by HDACI, which in turn can sustain the active state of Akt to attenuate the cellular apoptosis.

In addition to the interaction with the Akt/PKB pathway, Hsp70 also directly interacts with different proteins of the tightly regulated programmed cell death machinery thereby blocking the apoptotic process at distinct key points. For example, Hsp70 can inhibit the apoptotic cascade (Gotoh et al. 2004; Stankiewicz et al. 2005), decrease formation of the functional apoptosome complex (Beere et al. 2000; Saleh et al. 2000), prevent late caspase dependent events such as activation of cytosolic phospholipase A2 and changes in nuclear morphology, and protect cells from forced expression of caspase-3 (Jaattela et al. 1998). Moreover, Hsp70 inhibits c-Jun N-terminal kinase (JNK) mediated cell death by suppressing JNK phosphorylation either directly and/or through the upstream SEK (Stress-activated protein kinase (SAPK)/extracellular signal-regulated kinase (ERK) kinase) kinase (Mosser et al. 2000; Meriin et al. 1999; Volloch et al. 1999), and hampers TNF mediated apoptosis by inhibition of ASK-1 (Park et al. 2002).

Effect of HDACI on Gut and Lung in Hemorrhagic Shock

Hemorrhagic shock is characterized by tissue perfusion which is insufficient to meet the oxygen and nutrient demands of cells. Host response to hemorrhagic shock involves a coordinated expression of mediators that act both locally and systemically with profound effects on organ function. The accumulated evidences suggest that gut and lung, especially gut, represent important site(s) of immune mediator production and inflammation. Although two major hypotheses, gut-lymph-lung axis (Deitch 2001; Deitch et al. 2006) and gut-liver-lung axis (Peitzman et al. 1995; Thuijls et al. 2009), have brought much debate based on their different findings, it is clear that hemorrhagic shock is associated with intestinal ischemia which makes it a proinflammatory organ. For the fact that most of these mediators are produced by cells of the immune system, significant immunoregulatory actions occur within the gut-liver and/or lymph-lung axis and in peripheral immune sites (Fig. 11.2). It is well established that the gut plays a pivotal pathogenic role in the pathogenesis of SIRS and multi-organ dysfunction syndrome (MODS) (Deitch 1992; Hassoun et al. 2001).

Depending on the duration of hemorrhagic shock, the degree of intestinal mucosal disruption which is the result of ischemia, increases and plays the main role in bacterial translocation (Baker et al. 1988). It is also postulated that epithelial disruption occurs, not only due to ischemia, but also due to the effect of protease or free oxygen radicals in the epithelium (Deitch et al. 1990). Hemorrhagic shock appears to promote bacterial translocation by injuring the gut mucosa and impairing its barrier function.

It is clear that gut ischemia can cause lung injury, but how can it be prevented? We proposed that HDACI could protect lung from gut-originated damage. To test this proposal, the superior mesenteric artery (SMA) of rats was clamped for 60 min to induce ischemia and then released for reperfusion. Without any treatment, gut ischemia induced production of pro-inflammatory cytokines or prostaglandin-like compound such as IL-6, cytokine-induced neutrophil chemoattractant (CINC), 8-isoprostane in lung tissues, and increased neutrophil lung infiltration. However, treatment with VPA significantly reduced these mediators in lung tissues and improved survival in a rat model of ischemia and reperfusion (Kim et al. 2011). It is not clear how VPA protects gut from ischemic damage and prevents acute lung injury. However, based on our recent findings it is conceivable that VPA maintains gut barrier in part through stabilizing intestinal tight junctions (TJ) (Fig. 11.2).

The function of gut barrier is based on intestinal tight junctions. Encircling epithelial cells, the intestinal TJ is a region where the plasma membrane of epithelial cells forms a series of contacts

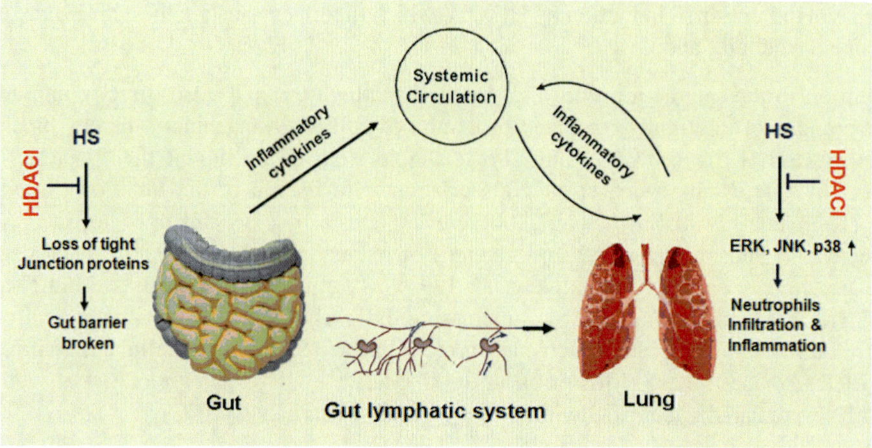

Fig. 11.2 Effect of HDACI on gut-lymph-lung axis in response to hemorrhagic shock. Hemorrhagic shock causes destruction of the gut barrier due to tight junction protein (e.g., claudin-3) loss. Bactria, endotoxin, and inflammatory cytokines enter into circulation and lung. In the lung tissue, MAPKs are stimulated and neutrophils infiltrated, resulting in acute lung injury. HDACI block these processing by inhibition of tight junction protein loss in gut and inactivation of MAPKs in lung

that appear to completely occlude the extracellular space and create an intercellular barrier and intra-membrane diffusion fence (Wong and Gumbiner 1997). Normally TJ is anchored in the cell via TJ proteins and filamentous actin (F-actin) cytoskeleton. Hypoperfusion, or ischemia, can cause disruption of F-actin cytoskeleton with subsequent TJ loss and barrier failure. Bacterial translocation to mesenteric lymph nodes, liver, and spleen is found at a very early stage of hemorrhagic shock (Thuijls et al. 2009). Loss of gut wall integrity not only leads to paracellular leakage of microbial products (Baker et al. 1988; Fink and Delude 2005), but also contributes to the development of systemic inflammation and distant organ failure (Hierholzer and Billiar 2001).

One of the major TJ proteins is claudin-3. Although the exact function of claudin-3 is not completely clear, it appears to be important in TJ formation and function (Morin 2005). Recent studies from our group (Li et al. 2010b) and others (Thuijls et al. 2009) have shown that HS leads to destruction of the gut barrier due to TJ protein loss. Also it was found that the claudin-3 protein is released into circulation very early (30–60 min) after the onset of HS (Li et al. 2010b). Alternatively, CINC, a chemokine that promotes neutrophil chemotaxis, is significantly elevated in serum and lung tissue with increased myeloperoxidase (MPO) levels at 4 h after hemorrhage. However, VPA treatment reversed HS-induced claudin-3 loss from the intestine and reduced the levels of CINC and MPO in serum and lung significantly (Fukudome et al. 2011). These findings suggest that VPA can stabilize claudin-3 in gut TJ, maintain the intestinal barrier, and prevent harmful gut-derived substances from getting into the systemic circulation.

Furthermore, we recently demonstrated that hemorrhagic shock (40% blood loss) resulted in phosphorylation (activation) of ERK, JNK and p38 mitogen-activated protein kinase (MAPK) in lung tissues at 1 h and 4 h. Post-shock administration of VPA (300 mg/kg, iv) significantly attenuated the MAPK activation without altering the total ERK, JNK and p38 proteins (Kochenek et al. 2010). These kinases are globally expressed and known to be key regulators of stress-mediated cell fate decision. Activation of these proteins has been strongly associated with poor outcomes while inhibition of MAP kinases has been associated with survival in hemorrhage models (Kochenek et al. 2010). VPA can also directly modulate MAPK activation. Cao et al. have studied the effects of HDACI treatment on LPS-induced activation of p38 MAPK and found that HDACI inhibit p38 phosphorylation. In their experiments, HDACI induce acetylation of MAP kinase phosphatase -1

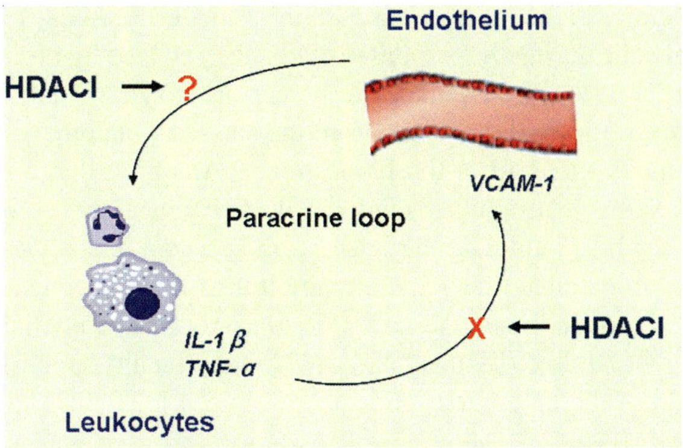

Fig. 11.3 HDACI modulate a paracrine between leukocytes and endothelial cells. The interaction between leukocytes and endothelial cells seems to form a paracrine loop. On one hand, IL-1β and TNF-α produced by infiltrating inflammatory cells such as neutrophils and macrophages can induce endothelial cells to express cytokines and adhesion molecules. On the other hand, endothelial cells play an essential role in speeding up this process via their ability to express cell surface adhesion molecules that mediate interactions with leukocytes in the bloodstream. HDACI break the loop by suppressing TNF-α induced VCAM-1 expression and reducing immune cells adhesion to endothelial cells

(MKP-1), a protein that dephoaphorylates MAPK and inactivates MAPK pathways. These results demonstrate that HDACI treatment and MKP-1 acetylation increases the interaction between MKP-1 and p38 MAPK, and results in p38 inactivation, reduced inflammation and increased survival among LPS-exposed mice (Cao et al. 2008). Whether similar acetylation-mediated mechanisms exist for the regulation of ERK and JNK is still unknown but is highly plausible.

Breakage of a Paracrine Loop Between Leukocytes and Endothelial Cells

The development of MODS is a critical problem in patients who suffer major blood loss. Although the precise mechanisms and pathways leading to organ injury after HS are still incompletely understood, neutrophils are thought to be a principal mediator of tissue damage. Migration of neutrophils into the tissues during HS leads to significant organ damage through the release of proteases and oxygen-derived radicals (Weiss 1989). Intravital microscopy studies have established a sequence of events involved in leukocyte migration to extravacscular sites: rolling, firm adhesion, and trans-endothelial migration. Under conditions of flow, leukocytes are first observed to roll along the endothelium of postcapillary venules. Subsequently, some of the rolling leukocytes stick firmly, migrate along the endothelial surface, diapedese between endothelial junctions, and then migrate through the sub-endothelial matrix (Harlan and Winn 2002).

The interaction between leukocytes and endothelial cells seems to form a paracrine loop (Fig. 11.3), and is instrumental in the migration of neutrophils into different tissues. On one hand, leukocyte recruitment to endothelium is a fundamental element of the tissue response to HS. Interleukin (IL)-1 and tumor necrosis factor-α (TNF-α) produced by infiltrating inflammatory cells such as neutrophils and macrophages can induce endothelial cells to express cytokines as well as adhesion molecules (Vadlamani and Lyengar 2004). On the other hand, endothelial cells play an essential role in speeding up this process via their ability to express cell surface adhesion molecules that mediate interactions with leukocytes in the bloodstream. Activated endothelial cells express molecules involved in leukocyte rolling, such as P- and E-selectin, leukocyte adhesion molecules including vascular cell adhesion molecule-1 (VCAM-1) and intercellular adhesion molecule-1 (ICAM-1), as well as chemoattractants such as chemokine (c-c motif) ligand 2 (CCL2) and IL-8, which can induce arrest of rolling leukocytes and promote leukocyte emigration from the vasculature (Petri et al. 2008; Zarbock and Ley 2009). Activation of endothelial cells during the response is typically induced by proinflammatory cytokines, such as TNF-α and IL-1β, released from leukocytes in response to HS (Cheng et al. 2010). Expression of adhesion molecules and additional cytokines from the activated endothelial cells in turn facilitates migration of additional neutrophils from the bloodstream. Anything that breaks this vicious circle could potentially attenuate the neutrophils mediated cellular damage.

HDACI have been shown to modulate the interaction between leukocytes and endothelial cells. Studies of Inous et al. demonstrated that TSA, an inhibitor of HDAC, can suppress TNF-α induced VCAM-1 expression and reduce monocyte adhesion not only to endothelial cells in vitro but also to venules in inflamed mice in vivo (Inoue et al. 2006). These results suggest that HDACI might be useful for the attenuation of the deleterious cross-talk between leukocytes and endothelial cells.

HDACI in Models of Septic Shock

The progression of infection to septic shock begins with the release of inflammatory mediators at the local site of microbial invasion. This induces the migration of white blood cells and platelets to the infection site and contributes to endothelial damage and increased micro-vascular permeability. Blood flow is also reduced which sets the stage for ischemia-reperfusion injury. These physiologic processes are part of the exaggerated SIRS, which can lead to MODS.

Many of the body's reactions to infection with gram-negative organisms are due to LPS, a component of the outer bacterial cell wall membrane, also referred to as endotoxin. It can induce septic shock physiology and has been used extensively to produce shock in laboratory models. LPS exerts the downstream signals through the Toll Like Receptor-4 (TLR4). TLR4 activates two downstream pathways: Myeloid differentiation factor 88 (MyD88)-dependent and MyD88-independent pathways. The former one leads to the production of proinflammatory cytokines such as IL-6, TNF-α and IL-12 with the quick activation of nuclear factor-kappaB (NF-kB) and MAPK. The MyD88-independent pathway is associated with the activation of interferon (IFN) regulatory factor 3 (IRF3), subsequent induction of IFN-β, and maturation of dendritic cells.

It has been shown that HDACI exert anti-inflammatory activities via the suppression of inflammatory cytokines and nitric oxide (Blanchard and Chipoy 2005). In LPS-stimulated human peripheral blood mononuclear cells, ITF2357 (an inhibitor of HDAC) reduces the release of TNF-α, IL-1β and IFN-γ (Leoni et al. 2005). Other HDACI such as TSA and SAHA have been shown to decrease LPS-induced inflammation in mice (Cao et al. 2008; Li et al. 2009; Li et al. 2010a). In RAW 264.7 cells, treatment of the macrophages with SAHA significantly suppresses LPS-induced gene expression and protein production of IL-1β, IL-6, and TNF-α (Li et al. 2009; Li et al. 2010a). In an in vivo rodent model of septic shock, HDACI attenuate acute lung and liver injury, and improve survival (Zhang et al. 2009; Li et al. 2010a; Zhang et al. 2010). Further mechanistic studies have demonstrated that HDACI play an important inhibitory role in TLR-4-MyD88 signaling pathways via NF-kB and MAPKs. Thus, protein acetylation is a potential regulatory mechanism that can modulate the inflammatory response (Fig. 11.4).

HDACI Affect NF-kB Activity

NF-kB is an ubiquitiously expressed transcription factor that plays an important role in innate immunity and other critical processes. The NF-kB family consists of p50, p52, p65 (Rel A), c-Rel and Rel B, which form homo- or hetero-dimers. The p50/p65 heterodimer is the most frequently found combination in mammals. Inactive NF-kB complexes are retained in the cytoplasm by the

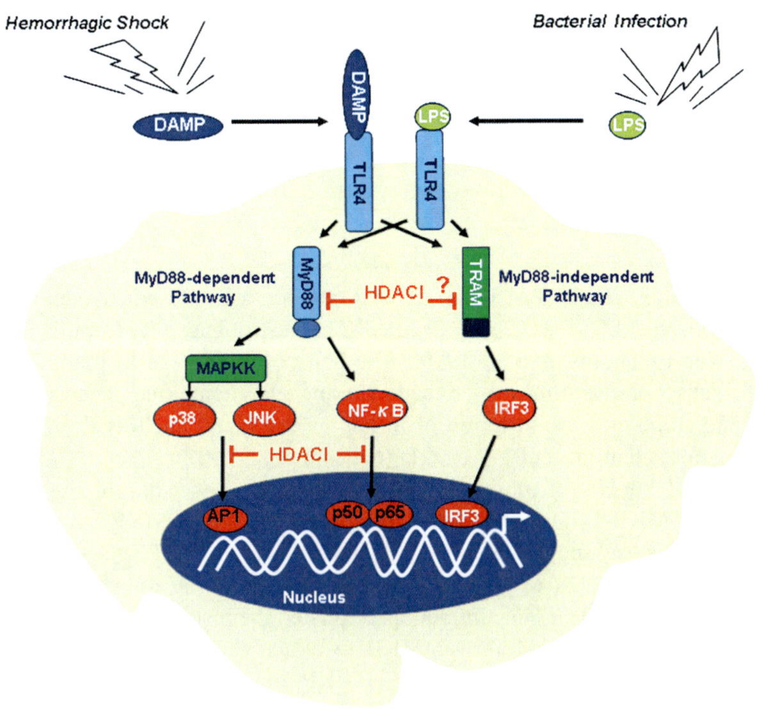

Fig. 11.4 TLR4 signaling – a converged immune response pathway for hemorrhagic shock and septic shock. TLR4 not only serves as a key sensor of pathogen-associated molecular patterns (PAMPs), but also is proposed recently to act as a receptor for some endogenous molecules called "alarmins". HDACI block TLR4 signaling at multiple steps; therefore they can inhibit immune response for both hemorrhagic shock and septic shock

IkB inhibitor. In innate immune signaling, host cells can respond to the threat of bacterial pathogens (e.g., LPS) via extracellular receptor TLRs (e.g., TLR4). TLR4 interacts with MyD88 and recruits interleukin-1 receptor-associated kinase 1 (IRAK1) and IRAK4 to the receptor complex. IRAK phosphorylates TNF receptor associated factor 6 (TRAF6) leading to the activation of the IkB kinase (IKK). The activation of IKK results in IkB phosphorylation, triggering its ubiquitination and proteasomal degradation. Free NF-kB then translocates to the nucleus to regulate the transcription of chemokines, cytokines, and other inflammatory response molecules (Hayden and Ghosh 2008).

In the nucleus, p50 and p60 can be regulated by acetylation. The function of acetylated NF-kB is complicated. Acetylation of p50 at K431, K440 and K441 promotes higher DNA binding affinity towards NF-kB target sequences correlating with increased p300 (histone acetyltransferase) recruitment and transcriptional activation (Deng and Wu 2003; Chen and Greene 2004). P300 can acetylate p65 at multiple lysine residues and result in different consequences. Acetylation of p65 at K221 and K310 is associated with an increased transcription of NF-kB target genes (Chen and Greene 2004) and is required for the full activity of p65 (Chen et al. 2002). In contrast, HDAC1 and HDAC3 deacetylate p65 at either K221 or K310, resulting in the inhibition of NF-kB. Additionally, K122 and K123 acetylation reduces p65 DNA binding affinity accompanied with increased IkB interaction and nuclear export (Kiernan et al. 2003). The p300-mediated acetylation of K314 and K315 in p65 has no obvious effect on NF-kB DNA binding or translocation.

HDACI have been shown to induce hyperacetylation and repress NF-kB signaling and expression of several target genes (Huang et al. 1997; Inan et al. 2000; Kramer et al. 2001). Conversely, other group reported that HDACI enhance NF-kB dependent gene expression in the presence of TNF-α (Adam et al. 2003; Ashburner et al. 2001; Quivy et al. 2002; Vanden Berghe et al. 1999). Presumably, inhibitory or enhancive effects of HDACI on NF-kB rely on the cell type and expression of a different set of HDAC isoforms as well as the source of cell stimulation (e.g. LPS, cytokines and high glucose levels) (Blanchard and Chipoy 2005).

HDACI Inhibit MAPK Activity

In mammalian cells, JNK and p38 MAPKs activate mitogen and stress-activated protein kinase 1 (MSK1) such as ribosomal S6 kinase 2 (RSK2). RSK2 has a strong activity towards phosphorylation of histone H3 at Ser10 (Thomson et al. 1999). The phosphorylation of histone H3 occurs on the promoters of the subset on the stimulus-induced cytokine and chemokine genes, recruits NF-kB to the promoters, and stimulates transcription of inflammatory genes such as IL-6, IL-8, IL-12 and macrophage chemoattractant protein 1 (MCP-1) (Saccani et al. 2002).

It has been reported that HDAC inhibitor TSA enhances the activity of mitogen-activated protein kinase phosphotase-1 (MKP-1) (Chi and Flavell 2008; Cao et al. 2008). MKP-1 is a nuclear-localized dual-specificity phosphatase and preferentially dephosphorylates MAPKs such as p38 and JNK. Recently, Cao et al. showed that MKP-1 interacts with HAT, and acetylation of MKP-1 inhibits TLR4 signaling (Cao et al. 2008). They immunoprecipitated the histone acetylase p300 and showed that it was associated with MKP-1. Moreover, MKP-1 was acetylated by p300 on lysine residue K57 within its substrate-binding domain. Acetylation of MKP-1 induced by TSA enhanced the interaction between MKP-1 and p38 MAPK, suggesting that HDACI could increase the phosphatase activity and inactivate p38 MAPK. Indeed, TSA increased MKP-1 acetylation and blocked MAPK signaling in wild-type (WT) cells. However, TSA had no effect in cells lacking MKP-1. Furthermore, TSA reduced inflammation and mortality in WT mice treated with LPS, but failed to protect MKP-1 knockout mice. These findings suggest that acetylation of MKP-1 inhibits innate immune signaling, and targeting the MAPK pathway by HDACI may be an important approach in the treatment of septic shock.

Recently, our group has found that HDAC inhibitor SAHA can reduce the expression of MyD88 gene and protein in vitro and in vivo after LPS insult (Li et al. 2010a). Moreover, SAHA acetylates heat shock protein 90 (Hsp90) and de-associates the protein from IRAK1, resulting in IRAK1 degradation (Chong et al. 2011). Our new findings have provided evidence that inhibition of HDAC can block, at least in part, activity of NF-kB and MAPKs in the initial steps of the TLR4-MyD88-NF-kB/MAPK pathway (Fig. 11.4).

TLR4 Signaling – A Converged Immune Response Pathway for Hemorrhage and Sepsis

Hemorrhage and sepsis activate several inflammatory and innate immune signaling pathways (Murphy et al. 2004). Systemically, these pathways promote recruitment of neutrophils and release of inflammatory cytokines (Botha et al. 1995). Within the cells, the inflammatory stimuli induce MAPK-dependent phosphorylation or phosphoacetylation of histone proteins and modulate the epigenetic accessibility of DNA (Saccani et al. 2002). Downstream, these signals change the expression profile of the genes in different cells types, altering the competing signals (e.g. pro-survival and pro-death, or anti-inflammation and pro-inflammation) that ultimately determine their fate. Two key pathways in hemorrhage- and sepsis-induced cellular injuries are the mitogen activated protein (MAP) kinase and NF-kB pathways. These proteins, ERK1/2, JNK, p38 protein kinase, and NF-kB, are globally expressed and known to be key regulators of cell fate decisions (Winter-Vann and Johnson 2007; Perkins 2007), which are involved in TLR4 signaling pathway.

It is well known that mammalian TLR4 serves as a key sensor of pathogen-associated molecular patterns (PAMPs) such as LPS. More recently, an additional role for TLR4 has been proposed. A number of reports have emerged to suggest that diverse molecules of host-cell origin may also serve as endogenous ligands of TLR4 (Bianchi 2007; Mollen et al. 2008; Tsan and Gao 2009). These molecules represent members of a recently identified family of molecules including Hsp70, fibrinogen, high mobility group box 1 (HMGB1), nucleolin, and annexins, etc. (Bianchi 2007). They have been found to serve as mediators of inflammation that may be expressed or released in response to tissue damage from trauma including HS. These molecules have been described as "alarmins", which are the equivalent of PAMPs but are endogenous molecules. They are rapidly released following non-programmed cell death not by apoptotic cells. Immune cells can also be induced to produce and release alarmins without dying. Generally, this is done by using specialized secretion systems or by the endoplasmic reticulum (ER)-Golgi secretion pathway. Endogenous alarmins and exogenous PAMPs can be considered subgroups of a larger family of damage-associated molecular patterns (DAMPs). They convey a similar message and elicit similar responses through TLR4 (Fig. 11.4) leading to activation of MAPKs and NF-kB pathways (Bianchi 2007; Roger and Babensee 2010).

HDACI have been described above for their pro-survival and anti-inflammatory properties. The combined pro-survival and anti-inflammatory effectiveness makes them a highly attractive choice for the treatment of lethal hemorrhagic shock, and its septic complications. In our preliminary studies, we have already discovered that HDACI not only inhibit expression of pro-inflammatory cytokines and chemokines in cells, but also prevent some alarmins from being released from cells in hemorrhagic shock and septic shock (Li et al. 2011). Further investigation with different models (e.g., "two-hit" model) are being planned to further clarify the precise mechanisms of action and the role played by protein acetylation.

Future Studies: HDACI in a Model of Hemorrhagic and Subsequent Septic Shock (a Two-Hit Model)

Hemorrhage induces immunosuppression and enhances susceptibility to sepsis. This is evident from depression of lymphocyte functions, production of various lymphokines, macrophage expression of receptors involved in opsonin-mediated phagocytosis, and antigen presentation function of peritoneal, splenic, and Kupffer cells following hemorrhage. Depression in various immune functions is apparent immediately following hemorrhage and persists for a prolonged period of time despite volume resuscitation. It appears that the increased release of systemic mediators, such as IL-1, IL-6, TNF-α, is associated with marked depression in immune responses and increased susceptibility to infection following hemorrhage (Chaudry and Ayala 1993). TNF-α is the trigger cytokine released early after the onset of hemorrhage. This cytokine then activates various populations of macrophages to release IL-1, IL-6, TGF-β and others. Although studies have described additional cytokines following trauma that may also play an important role it appears that TNF-α, IL-1, IL-6, and TGF-β, may play pivotal regulatory roles in the sequence of events leading to protracted immunodepression following hemorrhage. It should be recognized that the cytokine cascade, or "cytokine storm," may be triggered and amplified in a complex and interdependent manner (Chaudry and Ayala 1993).

The immunosupressive process is now fairly well characterized. The aforementioned imbalance in acetylation of proteins seems to be involved in the setting of both HS and septic shock. However, it remains unclear how to reduce or prevent the subsequent severe sepsis after HS. Aggressive resuscitation, vasopressor therapy, nutritional support and antibiotic prophylaxis alone have been proven to be ineffective when trying to alter the cytokine expression and decrease the infection rate after HS (Esrig et al. 1977; Chaudry and Ayala 1993; Bauhofer et al. 2006).

Focusing on the cellular pathophysiology of hemorrhagic shock and septic shock, we have explored the strategy of using HDACI (with or without resuscitation) as protective agents. HDACI alter the acetylation status of proteins and therefore have the potential to modulate the genomic and proteomic changes induced by hemorrhage and sepsis. We have shown that HDACI can dramatically improve survival in lethal models of hemorrhagic shock in rat (Shults et al. 2008; Fukudome et al. 2010), and swine models (Alam et al. 2009). These inhibitors can protect cells from apoptosis and suppress expression of pro-inflammatory cytokines (Li et al. 2009; Fukudome et al. 2011). HDACI can improve survival in a rodent model of septic shock regardless of whether the inhibitors are administrated before or after LPS insult. Moreover, a recent study from our laboratory demonstrated that the HDAC inhibitor SAHA can normalize TNF-α levels following rat hemorrhage in vivo and LPS second hit in vitro (Sailhamer et al. 2008). All of results above indicate that modulation of protein acetylation by HDACI may be an effective therapy for hemorrhage-induced sepsis. However, a critical animal model is still needed to confirm it.

An animal model of the "two-hit" paradigm has been used to target protein molecules involved in hemorrhage-induced septic shock (Fan et al. 2000; Shih et al. 2003; Bauhofer et al. 2006). In this model, animals were subjected to a non-severe resuscitated HS followed by a small dose of intratracheal LPS (Fan et al. 2000), CLP (cecal ligation puncture, Shih et al. 2003), or PCI (peritoneal contamination and infection, Bauhofer et al. 2006). While neither the first (shock) nor the second (infection) insult alone induces severe injury, the combination causes lung neutrophil accumulation and increase albumin transpulmonary flux (Fan et al. 1998), organ injury, colony-forming units of microbes in lung and liver, and mortality. However, so far no in vivo two-hit studies have addressed the effects of HDACI on inflammatory responses caused by HS followed by the development of polymicrobial sepsis. Results from an ongoing experiment in our laboratory should fill this gap in the near future.

Special Consideration

Acetylation is rapidly emerging as a key mechanism that regulates the expression of numerous genes (epigenetic modulation through activation of nuclear histone proteins), functions of multiple non-histone proteins involved in key cellular functions such as cell survival, repair/healing, and inflammation. HDACI hold great promise as a new class of agents for restoration of protein acetylation that can play a therapeutic role in hemorrhagic and septic shock. However, some special consideration should be kept in mind when the mechanism of action for HDACI is analyzed.

Sex hormones significantly impact survival in models of hemorrhagic shock (Lomas-Niera et al., 2005). Findings of Choudhry et al. suggest that the depression of immune functions in trauma hemorrhage is severe in young males, and ovariectomized and aged females. In contrast, the immune functions in proestrus females following trauma-hemorrhage are maintained (Choudhry et al. 2007). Studies have also shown that the survival rate in proestrus females following trauma-hemorrhage and the induction of subsequent sepsis is significantly higher than in age-matched males and ovariectomized females. Furthermore, administration of female sex hormone 17β-estradiol in males and ovariectomized females after trauma-hemorrhage prevents the suppression of immune response. Recently, HDACI such as TSA and VPA have been found to induce estrogen response element transactivation. Estradiol treatment in turn can potentiate the HDACI effect (Suuronen et al. 2008). These results indicate that the interaction of sex hormones with HDACI could play a significant role in shaping the host response following trauma and hemorrhagic shock.

Cell type is a big factor when evaluating the impact of HDACI. The response of normal and transformed (e.g. neoplastic) cells to a given HDACI is often completely different. HDACI- induced phenotypes changes in transformed cells include growth arrest, activation of the extrinsic and/or intrinsic apoptotic pathways, autophagic cell death, reactive oxygen species (ROS)-induced cell death, mitotic cell death and senescence. In comparison, normal cells are relatively more resistant to HDACI-induced cell death (Kelly and Marks 2005; Dokmanovic et al. 2007). In many transformed cells, ROS-oxidation-reduction pathways are important mechanisms of HDACI-induced transformed cell death (Ungerstedt et al. 2005). Thioredoxin (Trx) acts as a hydrogen donor required for the activation of many proteins, including ribonucleotide reductase which is essential for DNA synthesis and transcription factors and is an antioxidation scavenger of ROS (Lillig and Holmgren 2007). HDACI upregulate the expression of Trx binding protein 2 (TBP2) (Butler et al. 2002), which binds and inhibits Trx activity and can cause downregulation of Trx in transformed but not normal cells (Butler et al. 2002; Ungerstedt et al. 2005). Trx is an inhibitor of apoptosis signal regulating kinase 1 (ASK1) (Saitoh et al. 1998). Therefore, inhibition of Trx by HDACI in the transformed cells subsequently results in cell apoptosis.

Another potential confounder in animal models is the administration of heparin. The anti-coagulant heparin is frequently used to maintain blood flow through catheters (Rana et al. 1992) and to anti-coagulate shed blood reserved for return during resuscitation. However, high doses of heparin have been associated with effects that can confound results in experimental models of HS (Lomas-Niera et al. 2005). For example, direct addition of heparin to cells has been found to reduce histone acetylation (Buczek-Thomas et al. 2008).

As for the "two-hit" model, the timing of a secondary insult is critical to the ultimate response of the host (Lomas-Neira et al. 2005). Lomas-Neira et al. reported that hemorrhagic shock (priming stimulus) followed 24 h by the induction of sepsis (triggering stimulus) produced significantly higher levels of pro-inflammatory cytokine IL-6, MIP-2a and increased MPO activity in lung tissue. These hemorrhage-induced increases dissipated when sepsis was induced 72 h after hemorrhagic shock. It was found that there is a period of priming during the first 24 h following an initial traumatic inflammatory insult. The priming time keeps the secondary stimulus potential to trigger neutrophil mediated tissue or organ injury (Ayala et al. 2002). Therefore, the timing for drug administration would be a crucial variable in a "two-hit" model.

In summary, the effects of HDACI depend upon the host "cell context" which in turn influences acetylation or the interaction of HDACs with histone and non-histone proteins. Ideally, comprehensive consideration of host gender, cell type, the timing of insult and the nature of host/cell stimulation should be taken into account when effects of the inhibitor are examined in an experimental model. Lack of attention to these details can create an erroneous impression of contradictory results.

Conclusion and Perspectives

Experimental evidence has shown that treatment with HDACI increase endurance of animals subjected to lethal blood loss. The survival benefit is seen even when the drugs are administered post-insult and is reproducible in different species including large animal models of poly-trauma. Protective properties of HDACI are not limited to hemorrhagic shock, as it can also improve survival in LPS models of septic shock. Administration of HDACI modulates the immune system to create a favorable phenotype not only during the acute phase of hemorrhagic shock but also later when the septic complications are likely to occur. Repeated successes of HDACI in well designed animal models of hemorrhagic shock (small and large animals) and septic shock (pre- and post-shock treatments) suggest that modulation of protein acetylation is potentially a very useful strategy for the treatment of these critical diseases.

The future success of shock treatment might come from the implementation of pro-survival and anti-inflammatory strategy in proper animal and cellular models. More studies should further elucidate the function of individual HDAC isoforms in severe hemorrhage and inflammation and assess potential effects of HDACI on sepsis following hemorrhagic shock. Since individual HDAC isoforms have distinctive physiological functions, it is important to develop next generation of HDACI. The new HDACI could potentially target specific HDAC isoforms and presumably result in improved efficacy relative to the first generation pan inhibitors, such as SAHA and TSA, but with little adverse effects. In addition to being used as pro-survival agents for severe trauma hemorrhage or anti-inflammatory compounds for deteriorative sepsis, HDACI could be used as combined pro-survival and anti-inflammatory drugs to prevent hosts from sepsis and even to treat sepsis following hemorrhage. The potential success of HDACI in the two-hit model might provide insight toward the development of pharmacological agents for the treatment of the shock of hemorrhage and sepsis.

Acknowledgements Dr. Alam acknowledges grant support from the National Institutes of Health (RO1 GM084127), Defense Advanced Research Projects Agency (W911NF-06-1-0220), Office of Naval Research (N000140910378), and the US Army Medical Research Material Command (GRANTT00521959).

Abbreviations

ASK1	Apoptosis signal regulating kinase 1
BAD	Bcl-xl/Bcl-2 associated death promoter
Bcl-2	B-cell lymphoma 2
β_2-AR	Beta2-adrenergic receptor
BMP7	Bone morphogenetic protein 7
CASP	Colon ascendant stent peritonitis
CBP	Cyclic AMP (cAMP) response element binding protein (CREBP) binding protein
CCL2	Chemokine (C-C motif) ligand 2
CINC	Cytokine-induced neutrophil chemoattractant
CLP	Cecal ligation puncture
DAMPs	Damage-associated molecular patterns
DNA	Deoxyribonucleic acid
DUSP5	Dual specificity protein phosphatase 5
ELISA	Enzyme-linked immunosorbent assay
ER	Endoplasmic reticulum
ERK	Extracellular signal regulated kinase
F-actin	Filamentous actin
FWB	Fresh whole blood
GSK-3β	Glycogen synthase kinase-3β
h	Hour
H	Histone
HATs	Histone acetylases
HDA1	Histone deacetylase A1
HDACs	Histone deacetylases
HDACI	Histone deacetylase inhibitors
HMGB1	High mobility group box 1
HS	Hemorrhagic shock
Hsp 70	Heat shock protein 70
Hsp 90	Heat shock protein 90
ICAM-1	Intercellular adhesion molecule-1

IFN	Interferon
IGF-1	Insulin-like growth factor 1
IKK	IkB kinase
IL	Interleukin
IRAK 1	Interleukin-1 receptor associated kinase 1
IRF3	Interferon regulatory factor 3
IV	Intravenous (injection into a vein)
JNK	c-Jun N-terminal kinase
LPS	Lipopolysaccharide
MAGUK	Membrane-associated guanylate kinase
MAP	Mean arterial pressure
MAPK	Mitogen-activated protein kinase
MEF2	Myocyte enhancer factor 2
MKP-1	MAP kinase phosphatase 1
MMTV	Mouse mammary tumor virus
MODS	Multi-organ dysfunction syndrome
MPO	Myeloperoxidase
MSK1	Mitogen and stress-activated protein kinase 1
MyD88	Myeloid differentiation factor 88
NAD	Nicrotinamide adenine dinucleotide
NF-kB	Nuclear factor kappa B
PAMPs	Pathogen-associated molecular patterns
PCAF	p300/CREB-binding protein-associated factor
PCI	Peritoneal contamination and infection
p300	p300 histone acetyl transferase
PGC-1α	Peroxisome proliferator-activated receptor γ coactivator-1α
PI3K	Phosphoinositide 3 kinase
PIP2	Phosphatidylinositol 4,5-bisphosphate
PIP3	Phosphatidyl-inositol,3,4,5 triphosphate
PKB	Protein kinase B
PTEN	Phosphatase and tensin homolog
ROS	Reactive oxygen species
RSK2	Ribosomal S6 kinase 2
SAHA	Suberoylanilide hydroxamic acid
SEK	Stress-activated protein kinase (SAPK)/extracellular signal-regulated kinase (ERK) kinase
SIRS	Systemic inflammatory response syndrome
SIRT	Sirtuins
SMA	Superior mesenteric artery
RT-PCR	Reverse transcription polymerase chain reaction
TBP2	Trx binding protein 2
TFs	Transcription factors
TJ	Tight junction
TLR4	Toll-like receptor 4
TNF-α	Tumor necrosis factor α
TRAF6	TNF receptor associated factor 6
TRB3	Tribbles 3
Trx	Thioredoxin
TSA	Trichostatin A
VCAM-1	Vascular cell adhesion molecule-1

VPA Valproic acid
VSMCs Vascular smooth muscle cells
WT Wild type

References

Adam E, Quivy V, Bex F et al (2003) Potentiation of tumor necrosis factor-induced NF-kappa B activation by deacetylase inhibitors is associated with a delayed cytoplasmic reappearance of I kappa B alpha. Mol Cell Biol 23(17):6200–6209

Alam HB, Shuja F, Butt MU, Duggan M, Li Y, Zacharias N, Fukudome EY, Liu B, Demoya M, Velmahos GC (2009) Surviving blood loss without transfusion in a swine poly-trauma model. Surgery 146(2):325–333

Ashburner BP, Westerheide SD, Baldwin AS Jr (2001) The p65 (RelA) subunit of NF-kappaB interacts with the histone deacetylase (HDAC) corepressors HDAC1 and HDAC2 to negatively regulate gene expression. Mol Cell Biol 21(20):7065–7077

Avery J, Etzion S, DeBosch BJ et al (2010) TRB3 function in cardiac endoplasmic reticulum stress. Circ Res 106(9):1516–1523

Avila AM, Burnett BG, Taye AA et al (2007) Trichostatin A increases SMN expression and survival in amousemodel of spinal muscular atrophy. J Clin Invest 117:659–671

Ayala A, Chung CS, Lomas J et al (2002) Shock-induced neutrophil mediated priming for acute lung injury in mice: divergent effects of TLR-4 and TLR-4/FasL deficiency. Am J Pathol 161(6):2283–2294

Baker JW, Deitch EA, Li M et al (1988) Hemorrhagic shock induces bacterial translocation from the gut. J Trauma 28(7):896–906

Bauhofer A, Lorenz W, Kohlert F et al (2006) Granulocyte colony-stimulating factor prophylaxis improves survival and inflammation in a two-hit model of hemorrhage and sepsis. Crit Care Med 34(3):778–784

Beere HM, Wolf BB, Cain K et al (2000) Heat-shock protein 70 inhibits apoptosis by preventing recruitment of procaspase-9 to the Apaf-1 apoptosome. Nat Cell Biol 2(8):469–475

Bhatia M, He M, Zhang H et al (2009) Sepsis as a model of SIRS. Front Biosci 14:4703–4711

Bianchi ME (2007) DAMPs, PAMPs and alarmins: all we need to know about danger. J Leukoc Biol 81(1):1–5

Blanchard F, Chipoy C (2005) Histone deacetylase inhibitors: new drugs for the treatment of inflammatory diseases? Drug Discov Today 10(3):197–204

Botha AJ, Moore FA, Moore EE et al (1995) Early neutrophil sequestration after injury: a pathogenic mechanism for multiple organ failure. J Trauma 39(3):411–417

Buczek-Thomas JA, Hsia E, Rich CB et al (2008) Inhibition of histone acetyltransferase by glycosaminoglycans. J Cell Biochem 105(1):108–120

Butler LM, Zhou X, Xu WS et al (2002) The histone deacetylase inhibitor SAHA arrests cancer cell growth, up-regulates thioredoxin-binding protein-2, and down-regulates thioredoxin. Proc Natl Acad Sci USA 99(18):11700–11705

Camelo S, Iglesiuas AH, Hwang D et al (2005) Transcriptional therapy with the histone deacetylase inhibitor trichostatin A ameliorates experimental autoimmune encephalomyelitis. J Neuroimmunol 164:10–21

Cao W, Bao C, Padalko E, Lowenstein CJ (2008) Acetylation of mitogen-activated protein kinase phosphatase-1 inhibits Toll-like receptor signaling. J Exp Med 205(6):1491–1503

Carey N, La Thangue NB (2006) Histone deacetylase inhibitors: gathering pace. Curr Opin Pharmacol 6:369–375

Cha JH (2000) Transcriptional dysregulation in Huntington's disease. Trends Genet 23(9):387–392

Champion HR, Bellamy RF, Roberts CP, Leppaniemi A (2003) A profile of combat injury. J Trauma 5:S13–S19

Chang KT, Min KT (2002) Regulation of lifespan by histone deacetylase. Ageing Res Rev 1(3):313–326

Chang JG, Hsieh-Li HM, Jong YJ et al (2001) Treatment of spinal muscular atrophy by sodium butyrate. Proc Natl Acad Sci USA 98:9808–9813

Chaudry IH, Ayala A (1993) Mechanism of increased susceptibility to infection following hemorrhage. Am J Surg 165(2A Suppl):59S–67S

Chen LF, Greene WC (2004) Shaping the nuclear action of NF-kappaB. Nat Rev Mol Cell Biol 5(5):392–401

Chen LF, Wu Y, Greene WC (2002) Acetylation of RelA at discrete sites regulates distinct nuclear functions of NF-kB. EMBO J 21(23):6539–6548

Cheng Q, McKeown SJ, Santos L et al (2010) Macrophage migration inhibitory factor increases leukocyte-endothelial interactions in human endothelial cells via promotion of expression of adhesion molecules. J Immunol 185:1238–1247

Chesley A, Lundberg MS, Asai T, Xiao RP, Ohtani S, Lakatta EG, Crow MT (2000) The beta(2)-adrenergic receptor delivers an antiapoptotic signal to cardiac myocytes through G(i)-dependent coupling to phosphatidylinositol 3'-kinase. Circ Res 87:1172–1179

Chi H, Flavell RA (2008) Acetylation of MKP = 1 and the control of inflammation. Sci Signal 1(41):pe44
Choudhry MA, Bland KI, Chaudry IH (2007) Trauma and immune response-effect of gender differences. Injury 38(12):1382–1391
Chuang DM, Leng Y, Marinova Z, Kim HJ, Chiu CT (2009) Multiple roles of HDAC inhibition in neurodegenerative conditions. Trends Neurosci 32(11):591–601
Dangond F, Gullans SR (1998) Differential expression of human histone deacetylase mRNAs in response to immune cell apoptosis induction by trichostatin A and butyrate. Biochem Biophys Res Commun 247:833–837
de Ruijter AJ, van Gennip AH, Caron HN et al (2003) Histone deacetylases (HDACs): characterization of the classical HDAC family. Biochem J 370(Pt 3):737–749
DeBosch B, Sambandam N, Weinheimer C, Courtois M, Muslin AJ (2006) Akt2 regulates cardiac metabolism and cardiomyocyte survival. J Biol Chem 281:32841–32851
Deitch EA (1992) Multiple organ failure. Pathophysilogy and potential future therapy. Ann Surg 216:117–134
Deitch EA (2001) Role of the gut lymphatic system in multiple organ failure. Curr Opin Crit Care 7(2):92–98
Deitch EA, Bridges W, Ma L et al (1990) Hemorrhagic shock-induced bacterial translocation: the role of neutrophils and hydroxyl radicals. J Trauma 30:942–952
Deitch EA, Xu D, Kaise VL (2006) Role of the gut in the development of injury- and shock induced SIRS and MODS: the gut-lymph hypothesis, a review. Front Biosci 11:520–528
Deng WG, Wu KK (2003) Regulation of inducible nitric oxide synthase expression by p300 and p50 acetylation. J Immunol 171(12):6581–6588
Dokmanovic M, Clarke C, Marks PA (2007) Histone deacetylase inhibitors: overview and perspectives. Mol Cancer Res 5(10):981–989
Du K, Herzig S, Kulkarni RN et al (2003) TRB3: a tribbles homolog that inhibits Akt/PKB activation by insulin in liver. Science 300:1574–1577
Esrig BC, Frazee L, Stephenson SF et al (1977) The predisposition to infection following hemorrhagic shock. Surg Gynecol Obstet 144(6):915–917
Fan J, Marshall JC, Jimenez M et al (1998) Hemorrhagic shock primes for increased expression of cytokine-induced neutrophil chemoattractant in the lung: role in pulmonary inflammation following lipopolysaccharide. J Immunol 161:440–447
Fan J, Kapus A, Li YH et al (2000) Priming for enhanced alveolar fibrin deposition after hemorrhagic shock: role of tumor necrosis factor. Am J Respir Cell Mol Biol 22(4):412–421
Faraco G, Pancani T, Formentini L et al (2006) Pharmacological inhibition of histone deacetylases by suberoylanilide hydroxamic acid specifically alters gene expression and reduces ischemic injury in the mouse brain. Mol Pharmacol 70(6):1876–1884
Ferrante RJ, Kubilus JK, Lee J et al (2003) Histone deactylase inhibition by sodium butyrate chemotherapy ameliorates the neurodegenerative phenotype in Huntington's disease mice. J Neurosci 23:9418–9427
Fink MP, Delude RL (2005) Epithelial barrier dysfunction: a unifying theme to explain the pathogenesis of multiple organ dysfunction at the cellular level. Crit Care Clin 21(2):177–196
Fukudome EY, Kochanek AR, Li Y et al (2010) Pharmacologic resuscitation promotes survival and attenuates hemorrhage-induced activation of extracellular signal-regulated kinase ½. J Surg Res 163(1):118–126
Fukudome EY, Li Y, Kochanek AR (2011) Pharmacologic resuscitation decreases circulating CINC-1 levels and attenuates hemorrhage-induced acute lung injury. Surgery in press
Gao T, Newton AC (2002) The turn motif is a phosphorylation switch that regulates the binding of Hsp70 to protein kinase C. J Biol Chem 277(35):31585–31592
Gonzales E, Chen H, Munuve R, Mehrani T, Britten-Webb J, Nadel A, Alam HB, Wherry D, Burris D, Koustova E (2006) Valproic acid prevents hemorrhage-associated lethality and affects the acetylation pattern of cardiac histones. Shock 25(4):395–401
Gonzales E, Chen H, Munuve RM, Mehrani T, Nadel A, Koustova E (2008) Hepatoprotection and lethality rescue by histone deacetylase inhibitor valproic acid in fatal hemorrhagic shock. J Trauma 65(3):554–565
Gotoh T, Terada K, Oyadomari S et al (2004) hsp70-DnaJ chaperone pair prevents nitric oxide- and CHOP-induced apoptosis by inhibiting translocation of Bax to mitochondria. Cell Death Differ 11(4):390–402
Granger A, Abdullah I, Huebner F, Stout A, Wang T, Huebner T, Epstein JA, Gruber PJ (2008) Histone deacetylase inhibition reduces myocardial ischemia-reperfusion injury in mice. FASEB J 22:3549–3560
Haggarty SJ, Koeller KM, Wong JC et al (2003) Domain-selective small-molecule inhibitor of histone deacetylase 6 (HDAC6)-mediated tubulin deacetylation. Proc Natl Acad Sci USA 100(8):4389–4394
Harlan JH, Winn RK (2002) Leukocyte-endothelial interactions: clinical trials of anti-adhesion therapy. Crit Care Med 30(5):S214–S219
Hassoun HT, Kone BC, Mercer DW et al (2001) Post-injury multiple organ failure: the role of the gut. Shock 15(1):1–10
Hayden MS, Ghosh S (2008) Shared principles in NF-kappaB signaling. Cell 132(3):344–362
Hierholzer C, Billiar TR (2001) Molecular mechanisms in the early phase of hemorrhagic shock. Langenbecks Arch Surg 386:302–308

Hockly E, Richon VM, Woodman B et al (2003) Suberoylanilide hydroxamic acid, a histone deacetylase inhibitor, ameliorates motor deficits in amousemodel of Huntington's disease. Proc Natl Acad Sci USA 100:2041–2046
Hu E, Dul E, Sung CM et al (2003) Identification of novel isoform-selective inhibitors within class I histone deacetylases. J Pharmacol Exp Ther 307(2):720–728
Huang N, Katz JP, Martin DR et al (1997) Inhibition of IL-8 gene expression in Caco-2 cells by compounds which induce histone hyperacetylation. Cytokine 9(1):27–36
Inan MS, Rasoulpour RJ, Yin L et al (2000) The luminal short-chain fatty acid butyrate modulates NF-kappaB activity in a human colonic epithelial cell line. Gastroenterology 118(4):724–734
Inoue K, Kobayashi M, Yano K et al (2006) Histone deacetylase inhibitor reduces monocyte adhesion to endothelium through the suppression of vascular cell adhesion molecule-1 expression. Arterioscler Thromb Vasc Biol 26(12):2652–2659
Jaattela M, Wissing D, Kokholm K et al (1998) Hsp70 exerts its anti-apoptotic function downstream of caspase-3-like proteases. EMBO J 17(21):6124–6134
Kelly WK, Marks PA (2005) Drug insight: Histone deacetylase inhibitors – development of the new targeted anticancer agent suberoylanilide hydroxamic acid. Nal Clin Pract Oncol 2(3):150–157
Kiernan R, Bres V, Ng RW et al (2003) Post-activation turn-off of NF-kappaB-dependent transcription is regulated by acetylation of p65. J Biol Chem 278(4):2758–2766
Kim AJ, Shi Y, Austin RC et al (2005) Valproate protects cells from ER stress-induced lipid accumulation and apoptosis by inhibiting glycogen synthase kinase-3. J Cell Sci 118:89–99
Kim KH, Oudit GY, Backx PH (2008) Erythropoietin protects against doxorubicin-induced cardiomyopathy via a phosphatidylinositol 3-kinase-dependent pathway. J Pharmacol Exp Ther 324:160–169
Kim K, Li Y, Jin G et al. (2011) Effect of valproic acid on acute lung injury in a rodent model of intestinal ischemia reperfusion. Resuscitation in press
Kishimoto H, Hamada K, Saunders M, Backman S, Sasaki T, Nakano T et al (2003) Physiological functions of Pten in mouse tissues. Cell Struct Funct 28:11–21
Kochenek AR, Fukudome EY, Eleanor JS et al (2010) Pharmacological resuscitation attenuates MAP kinase pathway activation and pulmonary inflammation following hemorrhagic shock in rodent model. Oral presentation at the annual meeting of the American college of surgeons, October 2010
Kramer OH, Gottlicher M, Heinzel T (2001) Histone deacetylase as a therapeutic target. Trends Endocrinol Metab 12(7):294–300
Krivoruchko A, Storey KB (2010) Epigenetics in anoxia tolerance: a role for histone deacetylase. Mol Cell Biochem 342(1–2):151–161
Kucharska A, Rushworth LK, Staples C et al (2009) Regulation of the inducible nuclear dual-specificity phosphatase DUSP5 by ERK MAPK. Cell Signal 21(12):1794–1805
Lane AA, Chabner BA (2009) Histone deacetylase inhibitors in cancer therapy. J Clin Oncol 27(32):5459–5468
Lee JO, Yang H, Georgescu MM et al (1999) Crystal structure of the PTEN tumor suppressor: implications for its phosphoinositide phosphatase activity and membrane association. Cell 99:323–334
Leoni F, Fossati G, Lewis EC et al (2005) The histone deacetylase inhibitor ITF2357 reduction of pro-inflammatory cytokines in vitro and systemic inflammation in vivo. Mol Med 11:1–15
Li Y, Liu B, Sailhamer EA, Yuan Z, Shults C, Velmahos GC, deMoya M, Shuja F, Butt MU, Alam HB (2008a) Cell protective mechanism of valproic acid in lethal hemorrhagic shock. Surgery 144(2):217–224
Li Y, Yuan Z, Liu B, Sailhamer EA, Shults C, Velmahos GC, Demoya M, Alam HB (2008b) Prevention of hypoxia-induced neuronal apoptosis through histone deacetylase inhibition. J Trauma 64(4):863–870
Li Y, Liu B, Zhao H, Sailhamer EA, Fukudome EY, Zhang X, Kheirbek T, Finkelstein RA, Velmahos GC, deMoya CA, Hales RA, Alam HB (2009) Protective effect of suberoylanilide hydroxamic acid against LPS-induced septic shock in rodents. Shock 32(5):517–523
Li Y, Liu B, Fukudome EY, Kochanek AR, Finkelstein R, Chong W, Jin G, Lu J, deMoya M, Velmahos GC, Alam HB (2010a) Survival lethal septic shock without fluid resuscitation in a rodent model. Surgery 148(2):246–254
Li Y, Liu B, Dillon S, Liu B et al (2010b) Identification of a novel potential biomarker in a model of hemorrhagic shock and valproic acid treatment. J Surg Res 159(1):474–481
Lillig CH, Holmgren A (2007) Thioredoxin and related molecules – from biology to health and disease. Antioxid Redox Signal 9(1):25–47
Lin T, Alam HB, Chen H, Britten-Webb J, Rhee P, Kirkpatrick J, Koustova E (2006) Cardiac histones are substrates of histone deacetylase activity in hemorrhagic shock and resuscitation. Surgery 139:365–376
Lin T, Chen H, Koustova E, Sailhamer EA et al (2007) Histone deacetylase as therapeutic target in a rodent model of hemorrhagic shock: effect of different resuscitation strategies on lung and liver. Surgery 141(6):784–794
Lomas-Niera JL, Perl M, Chung C et al (2005) Shock and hemorrhage: an overview of animal models. Shock 24(Suppl 1):33–39
Maehama T, Dixon JE (1998) The tumor suppressor, PTEN/MMAC1, dephosphorylates the lipid second messenger, phosphatidylinositol 3,4,5-trisphosphate. J Biol Chem 273:13375–13378
Mai A, Massa S, Pezzi R et al (2003) Discovery of (aryloxopropenyl) pyrrolyl hydroxyamides as selective inhibitors of class IIa histone deacetylase homologue HD1-A. J Med Chem 46(23):4826–4829

Marinova Z, Ren M, Wendland JR, Leng Y, Liang MH, Yasuda S, Leeds P, Chuang DM (2009) Valproic acid induces functional heat-shock protein 70 via Class I histone deacetylase inhibition in cortical neurons: a potential role of Sp1 acetylation. J Neurochem 111(4):976–987
Marks PA, Dokmanovic M (2005) Histone deacetylase inhibitors: discovery and development as anticancer agents. Expert Opin Investig Drugs 14:1497–1511
Marumo T, Hishikawa K, Yoshikawa M et al (2008) Epigenetic regulation of BMP7 in the regenerative response to ischemia. J Am Soc Nephrol 19:1311–1320
McCampbell A, Taye AA, Whitty L et al (2001) Histone deacetylase inhibitors reduce polyglutamine toxicity. Proc Natl Acad Sci USA 98:15179–15184
Meriin AB, Yaglom JA, Gabai VL et al (1999) Protein-damaging stresses activate c-Jun N-terminal kinase via inhibition of its dephosphorylation: a novel pathway controlled by HSP72. Mol Cell Biol 19(4):2547–2555
Mollen KP, Levy RM, Prince JM et al (2008) Systemic inflammation and end organ damage following trauma involves functional TLR4 signaling in both bone marrow-derived cells and parenchymal cells. J Leukoc Biol 83(1):80–88
Morin PJ (2005) Claudin proteins in human cancer: promising new targets for diagnosis and therapy. Cancer Res 65(21):9603–9606
Mosser DD, Caron AW, Bourget L et al (2000) The chaperone function of hsp70 is required for protection against stress-induced apoptosis. Mol Cell Biol 20(19):7146–7159
Moochhala S, Wu J, Lu J (2009) Hemorrhagic shock: an overview of animal models. Front Biosci 14:4631
Murphy TJ, Paterson HM, Mannick JA, Lederer JA (2004) Injury, sepsis, and the regulation of toll-like receptor responses. J Leukoc Biol 75(3):400–407
Okumura K, Mendoza M, Bachoo RM et al (2006) PCAF modulates PTEN activity. J Biol Chem 281:26562–26568
Oudit GY, Sun H, Kerfant BG, Crackower MA, Penninger JM, Backx PH (2004) The role of phosphoinositide-3 kinase and PTEN in cardiovascular physiology and disease. J Mol Cell Cardiol 37:449–471
Park HS, Cho SG, Kim CK et al (2002) Heat shock protein hsp72 is a negative regulator of apoptosis signal-regulating kinase 1. Mol Cell Biol 22(22):7721–7730
Peitzman AB, Billiar TR, Harbrecht BG et al (1995) Hemorrhagic shock. Curr Probl Surg 32(11):925–1002
Perkins ND (2007) Integrating cell-signaling pathways with NF-kappaB and IKK function. Nat Rev Mol Cell Biol 8(1):49–62
Petri S, Kiaei M, Kipiani K et al (2006) Additive neuroprotective effects of a histone deacetylase inhibitor and a catalytic antioxidant in a transgenic mouse model of amyotrophic lateral sclerosis. Neurobiol Dis 22:40–49
Petri B, Phillipson M, Kubes P (2008) The physiology of leukocyte recruitment: an in vivo perspective. J Immunol 180:6439–6446
Ping P, Murphy E (2000) Role of p38 mitogen-activated protein kinases in preconditioning: a detrimental factor or a protective kinase? Circ Res 86(9):989–997
Quivy V, Adam E, Collette Y et al (2002) Synergistic activation of human immunodeficiency virus type 1 promoter activity by NF-kappaB and inhibitors of deacetylases: potential perspectives for the development of therapeutic strategies. J Virol 76(21):11091–11103
Rana MW, Singh G, Wang P et al (1992) Protective effects of preheparinization on the microvasculature during and after hemorrhagic shock. J Trauma 32:420–426
Roger TH, Babensee JE (2010) Altered adherent leukocyte profile on biomaterials in toll-like receptor 4 deficient mice. Biomaterials 31(4):594–601
Ryu H, Lee J, Plpfsson BA et al (2003) Histone deacetylase inhibitors prevent oxidative neuronal death independent of expanded polyglutamine repeats via a Sp1-dependent pathway. Proc Nalt Acad Sci USA 100(7):4281–4286
Ryu H, Smith K, Camelo SI et al (2005) Sodium phenylbutyrate prolongs survival and regulates expression of anti-apoptotic genes in transgenic amyotrophic lateral sclerosis. J Neurochem 93:1087–1098
Saccani S, Pantano S, Natoli G (2002) p38-depedent marking of inflammatory genes for increased NF-kB recruitment. Nat Immunol 3(1):69–75
Saha RN, Pahan K (2006) HATs and HDACs in neurodegeneration: a tale of disconcerted acetylation homeostasis. Cell Death Differ 13:539–550
Sailhamer ES, Li Y, Smith EJ et al (2008) Acetylation: a novel method for modulation of the immune response following trauma/hemorrhage and inflammatory second hit in animals and humans. Surgery 144(2):204–216
Saitoh M, Nishitoh H, Fujii M et al (1998) Mammalian thioredoxin is a direct inhibitor of apoptosis signal-regulating kinase (ASK) 1. EMBO J 17:2596–26026
Saleh A, Srinivasula SM, Balkir L et al (2000) Negative regulation of the Apaf-1 apoptosome by Hsp70. Nat Cell Biol 2:476–483
Schwartzbauer G, Robbins J (2001) The tumor suppressor gene PTEN can regulate cardiac hypertrophy and survival. J Biol Chem 276:35786–35793
Scher MB, Vaquero A, Reinberg D (2007) SirT3 is a nuclear NAD+-dependent histone deacetylase that translocates to the mitochondria upon cellular stress. Genes Dev 21(8):920–928
Shih HC, Wei YH, Lee CH (2003) Magnolol alters cytokine response after hemorrhagic shock and increases survival in subsequent intraabdominal sepsis in rats. Shock 20(3):264–268

Shuja F, Tabbara M, Li Y, Liu B, Butt MU, Velmahos GC, deMoya M, Alam H (2009) Profound hypothermia decreases cardiac apoptosis through Akt survival pathway. J Am Coll Surg 209:89–99

Shults C, Sailhamer EA, Li Y, Liu B, Tabbara M, Butt MU, Shuja F, Demoya M, Velmahos G, Alam HB (2008) Surviving blood loss without fluid resuscitation. J Trauma 64(3):629–638

Stambolic V, Suzuki A, de la Pompa JL et al (1998) Negative regulation of PKB/Akt-dependent cell survival by the tumor suppressor PTEN. Cell 95:29–39

Stankiewicz AR, Lachapelle G, Foo CP et al (2005) Hsp70 inhibits heat-induced apoptosis upstream of mitochondria by preventing Bax translocation. J Biol Chem 280(46):38729–38739

Steffan JS, Bodai L, Pallos J et al (2001) Histone deacetylase inhibitors arrest polyglutamine-dependent neurodegeneration in Drosophila. Nature 413:739–743

Stiles B, Groszer M, Wang S et al (2004) PTENless means more. Dev Biol 273:175–184

St-pierre J, Drori S, Uldry M et al (2006) Suppression of reactive oxygen species and neurodegeneration by the PGC-1 transcriptional coactivators. Cell 127(2):397–408

Sugars KL, Rubinsztein DC (2003) Transcriptional abnormalities in Huntington disease. Trends Genet 19(5):233–238

Suuronen T, Ojala J, Hyttinen JM et al (2008) Regulation of ER alpha signaling pathway in neuronal HN10 cells: role of protein acetylation and Hsp90. Neurochem Res 33(9):1768–1775

Suzuki T (2009) Explorative study on isoform-selective histone deacetylase inhibitors. Chem Pharm Bull 57(9):897–906

Tamguney T, Strokoe D (2007) New insight into PTEN. J Cell Sci 120(Pt23):4071–4079

Thomson S, Clayton AL, Hazzalin CA et al (1999) The nucleosomal response associated with immediate-early gene induction is mediated via alternative MAP kinase cascades: MSK1 as a potential histone H3/HMG-14 kinase. EMBO J 18(17):4779–4793

Thuijls G, de Haan JJ, Derikx JP et al (2009) Intestinal cytoskeleton degradation precedes tight junction loss following hemorrhagic shock. Shock 31:164–169

Torella D, Rota M, Nurzynska D, Musso E, Monsen A, Shiraishi I et al (2004) Cardiac stem cell and myocyte aging, heart failure, and insulin-like growth factor-1 overexpression. Circ Res 94:514–524

Tsan MF, Gao B (2009) Heat shock proteins and immune system. J Leukoc Biol 85(6):905–910

Ungerstedt JS, Sowa Y, Xu WS et al (2005) Role of thioredoxin in the response of normal and transformed cells to histone deacetylase inhibitors. Pro Natl Acad Sci USA 102(3):673–678

Vadlamani L, Lyengar S (2004) Tumor necrosis factor alpha polymorphism in heart failure/cardiomyopathy. Congest Heart Fail 10(6):289–292

Valko M, Leibfritz D, Moncol J et al (2007) Free radicals and antioxidants in normal physiological functions and human disease. Int J Biochem Cell Biol 39(1):44–84

Vanden Berghe W, De Bosscher K, Boone E et al (1999) The nuclear factor-kappaB engages CBP/p300 and histone acetyltransferase activity for transcriptional activation of the interleukin-6 gene promoter. J Biol Chem 274(45):32091–32098

Vidali G, Gershey E, Allfrey VG (1968) Chemical studies of histone acetylation. The distribution of e-N-acetyllysine in calf thymus histones. J Biol Chem 243:6361–6366

Voelter-Mahlknecht S, Ho AD, Mahlknecht U (2005) Chromasomal organization and localization of novel class IV human histone deacetylase 11 gene. Int J Mol Med 16(4):589–598

Volloch V, Gabai VL, Rits S et al (1999) ATPase activity of the heat shock protein hsp72 is dispensable for its effects on dephosphorylation of stress kinase JNK and on heat-induced apoptosis. FEBS Lett 461(1–2):73–76

Weichert W (2009) HDAC expression and clinical prognosis in human malignancies. Cancer Lett 280(2):168–176

Weiss SJ (1989) Tissue destruction by neutrophils. N Engl J Med 320:365–376

Winter-Vann AM, Johnson GL (2007) Integrated activation of MAP3Ks balances cell fate in response to stress. J Cell Biochem 102(4):848–858

Wong V, Gumbiner BM (1997) A synthetic peptide corresponding to the extracellular domain of occluding perturbs the tight junction permeability barrier. J Cell Biol 136:399–409

Xu YX, Ayala A, Chaudry IH (1998) Prolonged immunodepression after trauma and hemorrhagic shock. J Trauma 44:335–341

Yao XH, Nyomba BL (2008) Hepatic resistance induced by prenatal alcohol exposure is associated with reduced PTEN and TRB3 acetylation in adult rat offspring. Am J Physiol Regul Integr Comp Physiol 294(6):R1797–R1806

Yellon DM, Downey JM (2003) Preconditioning the myocardium: from cellular physiology to clinical cardiology. Physiol Rev 83:1113–1151

Yildirim F, Gertz K, Kronenberg G et al (2008) Inhibition of histone deacetylation protects wildtype but not gelsolin-deficient mice from ischemic brain injury. Exp Neurol 210(2):531–542

Zacharias N, Sailhamer EA, Li Y, Liu B, Butt MU, Shuja F, Velmahos GC, de Moya M, Alam HB et al (2010) Histone deacetylase inhibitors prevent apoptosis following lethal hemorrhagic shock in rodent kidney cells. Resuscitation, 29 October 2010 [Epub ahead of print]

Zarbock A, Ley K (2009) Neutrophil adhesion and activation under flow. Microcirculation 16:31–42

Zhang L, Wan J, Jiang R et al (2009) Protective effects of trichostatin A on liver injury in septic mice. Hepatol Res 39:931–938
Zhang L, Jin S, Wang C et al (2010) Histone deacetylase inhibitors attenuate acute lung injury during cecal ligation and puncture-induced polymicrobial sepsis. World J Surg 34(7):1676–1683
Zhao TC, Cheng G, Zhang LX et al (2007) Inhibition of histone deacetylases triggers pharmacologic preconditioning effects against myocardial ischemic injury. Cardiovasc Res 76(3):473–481

Index

E. Mylonakis et al. (eds.), *Recent Advances on Model Hosts*,
Advances in Experimental Medicine and Biology 710, DOI 10.1007/978-1-4419-5638-5,